SCALE THEORY

SCALE THEORY

A Nondisciplinary Inquiry

Joshua DiCaglio

MINNESOTA

UNIVERSITY OF MINNESOTA PRESS
MINNEAPOLIS
LONDON

The University of Minnesota Press gratefully acknowledges the financial assistance provided for the publication of this book by the Department of English at Texas A&M University.

Portions of chapters 8 and 11 are adapted from "Scale Tricks and God Tricks, or The Power of Scale in *Powers of Ten*," in *Configurations* 28, no. 4 (Fall 2020): 459–90; copyright 2020 Johns Hopkins University Press and Society for Literature, Science, and the Arts.

Published by the University of Minnesota Press
111 Third Avenue South, Suite 290
Minneapolis, MN 55401-2520
http://www.upress.umn.edu

ISBN 978-1-5179-1206-2 (hc)
ISBN 978-1-5179-1207-9 (pb)

A Cataloging-in-Publication record for this book is available from the Library of Congress.

Printed in the United States of America on acid-free paper

UMP BmB 2021

Contents

Introduction

Learning to Scale

IN THE ATRIUM

On your way to the Hayden Planetarium in the New York Museum of Natural History, you may or may not notice an exhibit displayed in the atrium between the planetarium and a gift shop. This exhibit, titled *Scales of the Universe,* uses the sphere of the Hayden Planetarium to portray the relative size of the planets in the solar system. Along the walkway, a series of plaques show the major size scales of the universe, including a "powers of 10" diagram pointing to the relative sizes of objects from atoms to galaxies.

You may pause on this view for a moment and find this range of sizes interesting. But, most likely, you'll follow the crowd around to the planetarium entrance (or the gift shop), where the exciting science communication occurs. There, with incredible production value and, as of this writing, Neil deGrasse Tyson's dramatic narration, you'll learn about the far-flung objects of space and new, amazing discoveries about dark matter.

In this way, scale is often, in some sense, left in the atrium of science and human knowledge. We know that it is significant that we have discovered such magnitudes and this bewildering, nested nature of objects. But we get distracted by the main event, the reason we found these radically different scales to begin with: we want to know about these things we find at new scales, how they work, what they do, how they relate to each other. The production value there is much higher—it's where science feels like explanation, technologies find themselves intervening in helpful (and lucrative) ways, and medical advancements alleviate the human condition. We want to know about the vast or infinitesimal structure of the universe, but how long do we pause to consider the conditions for arriving at these different scales? We want to scale our businesses or intervene in the structures of matter, but rarely pause to consider what that reach into different scales implies and requires. We find ourselves sequencing, replicating, and even copyrighting genes, making statements about the possible toxicity of molecular components, and know that, somehow, the collective actions of *Homo sapiens* scratching, hacking, and fracking at the Earth are manifesting as ecological effects, but find ourselves struggling to describe how such scale domains work. We want to think globally and act locally but don't consider how we made that distinction in the first place.

Of course, plenty have highlighted scale—the *Scales of the Universe* exhibit exists alongside extensive scalar productions and descriptions. I point to this exhibit in the atrium merely to highlight how scale has been held as of secondary interest. We want to know how life works, how systems work, how information works, how galaxies form, how atoms get put together, how ecologies interact—and only then, because these require scale, do we find ourselves commenting on scale. The aim of this book is to remain as carefully and patiently as possible in the atrium and examine the contours of scale itself.

THE CENTRALITY OF SCALE

Once we turn to scale, it appears both essential and mystifying. As the ecologist Simon A. Levin once argued, scale is "the fundamental conceptual problem in ecology, if not in all of science."[1] This simple claim grows in significance as this seemingly specialized concern recurs at the heart of apparently disparate inquiries. We notice that all science moves within and along different scales: whether probing the relationship between quantum motions and stellar phenomena, between DNA molecules and the formation of bodies, or between human actions and ecological effects, scientists are forced to navigate bewildering scalar forms as they reveal elaborate scalar objects, map entangled scalar relations, and somehow delineate the sizes and locations of their sites of study.

The nonscientist may recognize something of scale in the familiar philosophical and political questions about parts and wholes, in which, as Levin continues, "dynamics at one level of organization can be understood as the collective behavior of aggregates." Science has brought into focus the question of parts and wholes according to discrete and empirically probed descriptive levels: how exactly *do* these atoms, cells, bodies, planets, solar systems, galaxies relate? All science must carefully and precisely specify the scales being considered. In doing so, science becomes inextricably entangled with this idea called "scale."

Why? Scale is required for specifying what is being seen and studied. It is, as Levin notes, fundamentally about observation: "When we observe the environment, we necessarily do so on only a limited range of scales; therefore, our perception of events provides us with only a low-dimensional slice of a high-dimensional cake."[2] Whether selected by choice or by a limitation in the organism, this slice changes what is observed. As a scientist, Levin is interested in how these different scales affect scientific practice. However, he notes that this observational problem is not limited to those organisms known as "scientists." Rather, "all organisms face the same dilemma: for particular life history stages, the realized environmental variability will be a consequence of the scales of experience." This statement is monumental. What we are able to see, do, react to, develop, and interact with (this "realized environmental variability") is not only *related* to scale but is a *consequence* of it. The scale of experience determines in some way what any organism—what we ourselves—is able to see and do.

Yet, these *Homo sapiens* move beyond this scale of experience as science's scalar descriptions saturate our ways of speaking and thinking and intervening. Levin's primary example of scalar methods is trying to understand the relation between "fine-scale phenomena" of human activity and changes in climate. If the problem of scale is a problem for science, how can it not also be a problem for all of us as we try to understand our relation to these cells and ecosystems, atoms and galaxies? Levin thus opens space for a larger claim: scale is not only a fundamental aspect of all science but is a fundamental aspect of all inquiry. Because scale underlies observation, it is implicit within our ways of dividing, grouping, and relating together the various aspects of the world. When we speak about society, culture, language, evolution, humanity, objects, matter, economics—we are working with certain assumptions about scale. Thus, we need a broader acknowledgment of scale's foundational and inescapable place in (re)defining our most basic concepts. As scale's relationship to science indicates, scale is a particular conceptual tool that has its own parameters, constraints, and rules. This book aims to outline these parameters.

OBSERVATION AND SCALE

Most experiences are limited to a set of objects and processes determined by the structure of these *Homo sapiens* bodies. However, our modes of inquiry and technological adjuncts have permitted us to extend perception into other scales, such that we must now define scale itself to orient ourselves to these new levels of observation.

Scale is not the level of observation, but a tool for establishing a reference point for domains of experience and interaction. Scale enters in the relation between two very different modes of experience. It shows how an entity not apparent in our perceptual field—for example, a carbon atom—fits within one's normal experience. Without scale, "atoms" would be hypothetical and abstract, as they were until atomic theory tied "atoms" to objects discernible at a particular scale deduced through scientific practice. When we embed the objects and processes of other scales into our concepts and language, scale becomes implicit or forgotten as the condition for speaking in this way even as we develop more complex descriptions about these objects. Do I notice, when I speak of "cells," that there is some implicit relationship between my body and these "cells"?

These transformations need to be examined so that we can trace out this elaborate structure of observation. We need a theory of scale, a means of looking at this way of looking.

A SCALAR VIEW

Look:

Looking at this body, I can think of these sensations as cellular signals and can muse on my emotions as neurotransmitter exchanges. I consider my microbiome and supplement my diet with probiotics while also carefully eradicating unwanted microbes in routine sanitizing. Fears arise about errant molecules delivered through plastics and pesticides. Other lively molecules called viruses hijack our globalized systems, uprooting and destroying lives and disrupting the world economy. For an ever-shrinking fee, I can have my DNA analyzed and face the odd relation between these spiraled molecules and some ambiguous data about this body's dispositions. I see images of cells, drawings of molecules, diagrams of atoms like infinitesimal solar systems, and know that somehow these computers are made from incredibly small etchings. I contemplate nanomachines and artificially produced microorganisms that might come to embody unknown intelligence; they show up in movies at least and seem more believable as technological devices embed themselves more deeply in my life.

Looking into the world, I can conceive of my very weight as the gravitational pull of the collective presence of a massive Earth that I routinely see pictured. I can look with unease at this superfluous packaging on my food that arrived from across the planet and contemplate the island of plastic gathering in the Pacific. I can pick up one of these electronic devices and immediately connect to locales around this planet or dive into a wealth of information that includes endless bits of people's lives spread out in an ever-increasing network. I can attempt to examine the political or economic climate—only to face uncertainty about who is acting, where, or to what ends (conspiracy theories proliferate). The simple numbers and narratives fed as signs of these global accumulations hardly serve to indicate what I know is the intricate dance of multinational corporations and globally spread governmental bodies carrying out extensive operations connecting together not just the unimaginably large 7.8 billion people on this planet but the immense ecological structure that supports that population. Advertisements remind me that somewhere people are starving and species are going extinct. The vastness of space only compounds the bewilderment; these stars and solar systems compounded into galaxies and groups of galaxies—how does this immense observable universe fit here in my world, this universe that I observe?

BEWILDERMENT AND POSTHUMAN EXPERIENCES

We are intimately familiar with scalar objects and descriptions, but once we begin to inquire into them we find them absolutely bewildering. This bewilderment is an inherent attribute of scale *because* it moves us beyond our human-centered ways of dividing and seeing reality. But scale performs this decentering by connecting human perception and concepts with that which exceeds them. I don't see cells as cells, the Earth as the Earth, atoms as atoms, or galaxies as galaxies. Only intervening modes of representation, navigated by a scalar notation, permit me to experience and understand the world in this way. Bewilderment arises as we attempt to reinterpret what we are experiencing in terms that exceed immediate experience. Yet, scale demands that *already within your everyday experiences* are other scales of being. Looking at this hand, you are already seeing cells, atoms, ecological relations, bits of stardust.

Thus, even though we all are familiar with scale, knowing of a variety of scalar objects, invoking scalar relations, and finding ourselves entangled in scalar productions, we must nonetheless learn to scale. Scale requires that we rethink our ways of describing and thinking about the cosmos according to new and counterintuitive rules. It is not enough to just speak of cells. It is not even enough to work as a scientist in a lab, with elaborate apparatuses, with cells. Until you have reoriented your whole conception of your body according to this conception of cells, you have not comprehended what scale is and what it implies. But scale does not stop there; it also demands that we comprehend the strangeness of the quantum, the vastness of a million galaxies, and the diffused intimacy of the ecological. And all of this, it says, is already implicit in your experience of gravity, in your relation to the sun, in the air you breathe.

SCALE AS A FUNDAMENTAL PHILOSOPHICAL QUESTION

This book traces out the conceptual, perceptual, and discursive aspects of scale writ large, including how we find ourselves with scale as a notion, the conditions under which it operates, and the difficulties it presents for human knowledge built largely on nonscalar experience. It tries, first and foremost, to highlight scale as a certain way of thinking and speaking, and to train the reader in this way of thinking. I hope to demonstrate that the longer you dwell with scale, the more fundamental it appears.

There are three foundational philosophical concepts that scale indelibly alters. First, scale deals with the delineation of objects, transforming one field of objects into others depending on the scale of inquiry. Scale suggests that any object appears different—or does not even exist at all—if you change the scale of perception. As two ecologists note, "If you move far enough across scales, the dominant processes change. It is not just that things get bigger or smaller, but the phenomena themselves change."[3] In a framework of scale, objects are no longer clearly separate, apparent, and self-contained entities "in themselves," since changes in scale reveal that the same thing also exists differently. All objects, therefore, must be defined via their scale of interaction and perception. In short, scale changes everything because it changes what every thing means.

Second, because scale deals with perspective, it is intertwined with notions of the subject, experience, consciousness, and interpretation. In applying scale to the body you call "mine" you find yourself looped into a disorienting question: whose body is it? The atoms'? The cells'? The planets'? The galaxies'? While this may seem absurd, this conundrum is already necessitated by scale and is, I argue, the most fundamental challenge presented by scale. We can only pretend that objects and relations exist "in themselves" and apart from these scalar delineations if we leave our conception of ourselves intact. Doing so makes scale inevitably abstract; it permits us to treat scalar objects as just any other objects and scalar relations as just more relations. But scale ties those relations back to our usual experience, saying that what I am is already atoms, cells, planets, galaxies. But then who is scaling?

Finally, scale reworks our notions of process, relation, and organization. Scale presents not only new objects but a tangle of new relations in two senses: first, scale reveals relationships that were previously not apparent simply because they were not discernible; and in turn, scale provides new possibilities for comprehending these aggregates and relations. Schemes of control and force—whether called "top-down," "bottom-up," the "great chain of being," or hierarchy of things—contain presumptions about scale that are revised when we carefully examine this apparatus.

The center of this book (Part II) directly addresses these reconfigurations, with a chapter devoted to each. But because scale entails a particular, counterintuitive way of thinking, we must first experiment with this way of thinking and attempt to make this term "scale" clear. Thus, Part I provides a series of thought experiments that define the parameters of scale as I am using it here and train the reader in thinking in terms of scale.

THE PHILOSOPHICAL STATUS OF SCALE

Before we begin these experiments, I want to highlight a few ways that scale has been positioned in theoretical conversations. In discussions of the "ontological status of scale," scholars tend to be critical of the concept partially because it is unclear what scale is.[4] Even in accounts favorable to scale, scale is said to be an "arbitrary mental device which allows us to make sense of our existence."[5] Or perhaps, "merely a way for humans to perceive and comprehend the world."[6] Or, as one geographer put it, "there is no such thing as scale."[7] Stripped of their diminutive qualities, these statements are partially correct: scale is a conceptual device that allows us to make sense of our existence, but not one which is entirely arbitrary. Scale provides a way for humans to perceive and comprehend the world, but one that is essential and not necessarily tied to humans. Finally, scale does not exist, in the sense that "scale" itself is not an entity but a reference for the structure of phenomena. The diminutive qualities of each of these statements relegates scale in relation to the social, the epistemological, and the ontological, respectively. But scale is not quite any of these.

To say that scale is ontological would imply that scale exists already as a kind of entity or fact of the world. To say that scale is epistemological would imply that it is a way of learning about or accessing the world. To say that scale is a social structure would imply that scale is one among many conceptual frameworks that society has built for mapping the world. And yet, scale relies both on the "being" of phenomena and a reference to the one who is measuring as a means of coming to know being. If we take the whole scaling process into account, considering whether scale is an epistemological, ontological, or social construct has less meaning. Rather, scale functions at a level above ontology and epistemology: scale is a means of orienting yourself both to experience and to the being of things. It is a metaphenomenal structure that makes use of a consistent measure to compare and calibrate ourselves to phenomena. Scale does not arise solely from beings (the ontological) or our mode of accessing beings (epistemology), but rather is a way of systematically organizing and comparing phenomena as they appear. Scale is an addendum and a reflecting back on what is discovered phenomenologically, first through attending to the presentation of the world and then systematically extending this presentation using a defined measure. In turn, the consistency of this defined measure then accounts for the way we are dividing reality, which means that scale cannot be solely a social construct.

Once discovered, scale becomes a necessary conceptual adjunct to any mode of inquiry—whether we want to speak of objects (ontology), our learning about objects (epistemology), or our way of organizing, speaking about, and distributing reality (sociality)—*since the objects in question are only understood through the reorientation mapped by scale.* Studying a cell only becomes possible once we extend our perceptual apparatus to sufficient resolution to identify cells, but it only becomes coherent when scale orients us to this perception. Regardless of the ontology of the cell (whether it actually exists), its epistemological status (how we find out about it), or the social aspects that go into its discovery (the training required to use a microscope), scale is required to understand what we mean by "cell."

Scale provides us with a reference point to take note of how any phenomenon compares to any other. Doing so permits us to systematically attend to how what is is revised and structured as beings dissolve, shift, combine, layer, and relate to each other differently depending on the scale of examination. Scale does not mediate our access to the world;

it permits us to make sense of mediated, extended, and projected experience. This status makes the scalar relation unlike any other kind of relation we find ourselves in. When we say that the Earth is made of rocks and the rocks are made of atoms, we are suggesting that the very being of each phenomenon is made up of, contained by, and holds reference to the others.

TWO KINDS OF TRANSFORMATION

With this outline, we can see how Levin's invocation of observation cannot be read within conceptions of objects, subjects, reality, interpretation, relations, and so on derived from this-scale experience. Rather, scale transforms the very possibilities of these terms. Scale requires that we move beyond the strict matters of organism and observation, matter and idea, language and knowledge. In short, scale forces us into what posthumanist theorists have recently emphasized as the deconstruction of the binaries of humanism, such as mind/matter, subject/object, and nature/human (see chapter 8). However, scale does not simply tell us to move beyond these binaries; it is a tool for moving us beyond them.

Demonstrably, Levin has done something quite interesting: in speaking of the "scale of observation," what has the scientist performed? On what scale is this observation taking place? Can any organism staying within its biological capacities of observation change scales sufficiently to observe that its observation relies on a scale of observation? In speaking of scale, Levin has transformed not only objects but his own level of observation.

In this way, scale navigates two vectors of transformation: it is at the intersection of a transformation of the world and a transformation of ourselves. Remaining at one scale, these observers and objects seem to solidify in a formation. But scale itself is the possibility of permitting these subjects and objects to transform again. In moving persistently across these transformations, scale further transforms the possible subjects and objects we might attribute to these words "world" and "ourselves." In many ways, this book is nothing more than an exegesis of the last four sentences.

Scale itself is properly transformative in turn, both changing and mapping our understanding of what we experience. Scale is inevitably rhetorical in the sense that it is about and requires a transformation or reconfiguration of our understanding of reality. As the study of persuasion, rhetoric focuses us on both the descriptive power of scale for objects and the transformative power of scale for those attending to these descriptions. Different rhetorical strategies are needed for us to follow out this transformation and permit scale to properly transform and reorient ourselves.

FINDING A THEORY OF SCALE ITSELF

Because we are so entangled in scale it is unsurprising that many *Homo sapiens* are already working through significant elements of scale, albeit obliquely or manifesting in other guises—subordinating scale to other issues, failing to follow through on the scalar concerns, or focusing on the scalar productions rather than the bewildering apparatus that makes them possible. Much recent work in critical theory and philosophy already relies on, or works with, scale. However, because they lack a clear theory of scale, studies that get the closest to describing scale often fall short of accounting for its most difficult implications.

One major problem is that it is difficult to study all scales at once. Just as Levin is primarily interested in scale for ecology, those studying science within the humanities often study a scale domain rather than scale itself. Much of the most significant work in science studies over the last few decades has focused on delineating and demonstrating the foreign nature of various scales. For example, Karen Barad's *Meeting the Universe Halfway* argues for revisions of agency and interpretation based on findings from the quantum scale. Work on nanotechnology demonstrates the strangeness of the nanoscale for conceptions of vision, control, and materiality.[8] The work of Natasha Myers, Stephan Helmreich, Myra Hird, and others treats in different ways the implications of the cellular and microbial scales.[9] Scholars of systems science and cybernetics, such as N. Katherine Hayles, Bruce Clarke, Cary Wolfe, Elizabeth Wilson, and Richard Doyle, have highlighted the entangled nature of systems thinking, particularly on the level of the planet.[10] Studies of space rhetoric and environmentalism, such as the work of Ursula Heise, Timothy Morton, and Lisa Messeri, have worked with the strange attempt to scale the seemingly abstract nature of place and outer space.[11] All of this work reveals an astonishing array of transformations in thought, culture, and language within these scalar shifts. All of these scholars inevitably confront their object of study as scalar, but these moments are, with some exceptions, subordinated to the primary scale of interest. Such work provides the basis for the following theory of scale, but one task here is to show how scale itself is a common thread in these discursive transformations.

Since I began this project, there has been a rapid increase of interest in scale in the humanities, particularly in the study of literature and culture.[12] However, such work tends to import theories of scale from the various disciplines (particularly geography) and handle several notions of scale together, while not making clear a theory of scale itself.[13] We thus want to develop more clearly a theory of scale itself that works across these diverse disciplines and objects. By "scale itself" I mean the apparatus of scale apart from any of the objects and processes that already make use of scale. I will show that there is a common notion of scale that needs to be accounted for by thinking not just *about* any one discipline or sets of objects but by thinking *across* them.

In scientific theory, many others aside from Levin have also directly engaged with scale and orders of magnitude, particularly in systems theory and hierarchy theory.[14] But these are illustrative of the constraints in approach that tend to come with scientific expertise—a tendency to focus on method that is limited or filtered through one's discipline (and the scales that one focuses on) and that remains largely confined to scientific practice and discourse. Thus, many of Levin's questions about scale are methodological (e.g., how does scale affect scientific practice?) and are read through ecology. Hierarchy theorists attempted to generalize their scalar concepts, but more work is needed to defend and clarify the nature of these scales, how we arrive at them, and what they imply.[15] Often attempts to do so, such as *The View from the Center of the Universe* by astronomer Joel R. Primack and science communicator Nancy Ellen Abrams, conceptualize what they are doing as primarily an act of science communication rather than a more significant, philosophical inquiry into how scale has made possible a different mode of understanding, thinking, and speaking about reality. In turn, Geoffrey West's *Scale,* a distillation of his scientific work written for a popular audience, extols scale as central to his work but does not contain a careful theory of scale, resulting in some significant omissions and confusions that we'll encounter later in this book. Finally, the field of geography has an extensive conversation on scale but, as we will discuss in chapter 9, remains largely confined to scales relevant to the human.

The task here is to think as generally as possible about as many of these disciplinary descriptions as possible and to extend these considerations out of the specialized concerns of science and into the more general questions about reality, knowledge, language, culture, and so on. To this end, a note about my own training, since it affects what follows, is in order. In many ways I have managed to avoid being disciplined. My training has been primarily in rhetoric, science studies, and critical theory of the sort that one might acquire in many English departments in the United States. This training has managed to keep me insulated somewhat from the usual disciplinarity, with my focus in rhetoric allowing me to skirt even the disciplinary limitations typical of literary studies. I treat rhetoric in a more philosophical sense as the study of the effects, constraints, and possibilities of communication. This notion of rhetoric provides a basis for this project. My approach has been to consider each of these sites as first and foremost an articulation attempting to make sense of the same thing: scale. These different ways of articulating scale have different effects on what it is and how we conceive it, but out of those discourses one can distill common elements. Once these common elements were highlighted, I was able to separate out the common issues and explore how these issues arise from or unfold out of scale. Thus, earlier drafts of this book took a form more typical of cultural studies. Each chapter began with a cultural depiction of scale and then treated this moment as representative of a key scalar trope out of which one could distill aspects of scale. This method resulted in extensive analysis of texts but spread the theory of scale throughout the text. This final version distills out the results of this abductive process by placing the conclusions at the beginning in a more systematic and theoretical manner in Part I. The results appear more deductive than intended. Thus, while this book makes many general conclusions, I consider it primarily an experiment in articulation for encountering the conditions of scale and the effects of scale on experience. Some reflections on this method (and its relationship to science studies and rhetoric) can be found in chapter 10.

The delineating of scale itself required that I consider a wide variety of disciplines, interests, and discourses. Perhaps such an approach should be called nondisciplinary, since the following work only considers disciplines insofar as they might serve as apparatuses for focusing certain methods, emphases, and articulations. I have often ignored disciplinary boundaries entirely. However, I have not left out ties and references to specialized conversations—even though they strain my personal expertise—since to do so would cut the project off from the many conversations equally important to the question of scale. I see such ties not so much as interventions but as a kind of ping (in the networking sense) to those whose specialized knowledge might more adequately respond. In doing so I always had to make a choice: dive more deeply and thoroughly into a specialized conversation, or stay focused on scale. The focus on scale itself is the guiding factor throughout the work. Where possible, to maintain rigor and clarity, I fall back onto the practice I am trained in most: careful analysis of a particular text regardless of its discipline and genre. Undoubtedly this results in many loose ends, omissions, and potential misunderstandings (although I have attempted to minimize the latter, particularly in relation to scientific facts and practice), but such is inevitable with a project of this magnitude.

Given my training, many of the interventions provided will be couched in conversations in critical theory and science studies, but my hope is that readers from many fields and forms of expertise will find the content here useful, assuming they are willing to forgive the ways that the approach differs from their modes of thinking or doing scholarship. To this end, Part I is written in the most axiomatic form, consisting of six thought experiments out of which various principles, parameters, and conundrums of scale are defined. The first three thought experiments, which I call "experiential origins of scale," culminate in a definition of scale (3.44—I will use this notation to cross-reference Part I with the remainder of the book), at which point the reader might want to move on to Part II. The final three chapters in Part I experiment further with scalar thinking, now assuming the apparatus of scale. By the end of Part I the reader will be adequately exposed to and trained in the conception of scale that is the core of this project.

Part II is equally philosophical, but now permits cultural depictions and critical engagements to enter as central to the arguments. This section examines more directly how scale reconfigures objects, subjects, and relations, but not exclusively as a philosophical problem. That is, I discuss basic philosophical problems not simply as abstract concepts but as a struggle to grasp and articulate a scalar configuration of reality. The arguments here will retrace and extend some of what was stated more axiomatically in Part I for the sake of something like prolepsis (the rhetorical term for accounting in advance for objections): how do we resist or struggle with what scale implies about subjects, objects, and relations?

In Part III, I consider directly the relationship between scale and communication, rhetoric, and representation. Since scale exceeds human experience, scalar descriptions must somehow depict or present that which is not immediately before us. What, then, does science describe? How do we interpret or approach scalar depictions? How do we make sense of scale's transformation of ourselves such that it actually changes our way of being and living in the world?

It is my hope that this book will facilitate a more extensive and direct conversation about scale that more adequately crosses disciplines, while providing a relatively comprehensive, but by no means exhaustive, outline of what it involves. The primary goal here is to bring attention to the most bizarre aspects of scale while also training the reader in what scalar thinking and speaking might entail. To that end, the reader can get a general sense of the book in much less time by reading the first three chapters, then the first two or three sections of each chapter of Parts II and III, which set up the chapter's problem and an overview of the way a theory of scale alters that problem.

MYSTICISM AND SCALE

A final note before we proceed. Perhaps the most controversial portion of this project is going to be its emphasis on a set of philosophical positions, practices, and discourses that may be labeled mysticism, the contemplative, or the nondual. It is necessary to situate this emphasis, since it is easily misunderstood.

There is a significant omission in our usual way of considering scale. The disorientation provided by scale and the fact that it relates to a transformation of ourselves means that it launches us necessarily into territory that is less comfortably scientific. This point of bewilderment is where those invested in scientific discourse tend to retreat back to the concrete productions of scientific study. We consider the cellular diagram rather than that my body is made up of cells; we return to studying images of stars and avoid contemplating our relationship to these vast distances; we reinscribe quantum fields as particles rather than consider the interpretive implications of quantum uncertainty. Again and again, the products of scale tempt us to retreat in this way, in part, because to do otherwise just seems too *mystical*. Thus, even Primack and Abrams, in one of their most interesting moments of scalar experimentation, throw out the phrase "It is not mystical; it is practical."[16] But what if scale *is* mystical? The problem is that we don't understand why it's mystical or what this means because most invocations of that term do not carefully trace out the relationship between scale and the history, practices, and rhetoric of mysticism.

I argue that mysticism describes a particular experience, particular practices designed around that experience, and a particular configuration toward reality *that arises from and works out a scalar perspective.* When scientists take the time to dwell on the larger implications and effects of scale, they find themselves entering the discursive realm of mysticism. In turn, those who have mystical experiences often find that the scalar descriptions of science provide a language that fits their experience. What if, to fully orient ourselves to scale, we have to set aside this repeated dismissal of mysticism? I will show how scale brings some surprising clarity to what mystics have always been going on about and, in turn, that mysticism helps clarify essential difficulties presented by scale.

To this end, mysticism can be defined in scalar terms as that branch of inquiry that aims to induce and integrate encounters with nonhuman scales, particularly the cosmic scale.[17] Mysticism arises in the experience of an existence larger than oneself, in the experience of experience exceeding itself. This glimpse of a vaster totality beyond human bounds reorients the "individual" outside of the typical scale of the human being. This person must then try to make sense of, describe, and develop this new orientation. Doing so often leads to the articulation of a scalar terminology; we'll encounter throughout this work the fact that most scalar articulations arise first within descriptions of contemplative experience.

The terms *mysticism, contemplative,* and *nondual* all bear a relationship to these scalar aspects. (We could add many other terms that will be caught up in the discussion, although we will not have space to address the nuances for each: *holism, idealism, monism, organicism, transcendentalism, perennialism,* and many more.) *Mysticism* describes a relationship to articulation: conceptual articulation is insufficient for the transformation experienced and described. Such articulations are "mysterious" because they are outside of the typical human lifeworld. *Contemplative* points to practices for cultivating this transformative experience. In turn, *nondual* describes a philosophical position, usually through a translation of the Sanskrit *advaita* (more on this in chapter 7). The scholar of religion Arthur Versluis gathers all three aspects together when he defines mysticism as "the contemplative ascent of the individual from duality to subject-object transcendence. . . . [The] transcendent is one, unitary, existing outside all time, abiding in eternity, unbounded, and immeasurable. Mysticism is a word for coming to recognize transcendent reality for oneself."[18]

The theory of scale presented here stumbles upon this same articulation, while clarifying this definition. The purpose of this project was not initially to comment on mysticism, but once the relationship between scale and mysticism became apparent it became essential to follow out these connections. It is crucial that the reader understand that these connections emerge outside of any dogmatic or religious context and are in many (but not all) senses secular. The approach here is again rhetorical: what are the functions of these terms, tropes, discourses, and practices of mysticism for describing and integrating in the scalar transformation? Ultimately, mysticism helps clarify the rhetorical elements of scale by accounting for how scale transforms the one who scales. For this argument to be made, we have to rethink mysticism in relation to what we have found out about scale and take it seriously as a discourse alongside science, philosophy, and social theory. However, because, I would argue, *mysticism is what is missing from our current conceptions of scale,* what follows will often counterpose nondual, contemplative conclusions to what can be drawn from many other discursive modes in order to highlight the difficulty and the task that scale has thrown us into.

The conclusions presented here are ways of seeing and taking into account the difficulty that our increasingly scaled understanding of reality presents for our persistently human-bound concepts. As an essential aspect of the structures of observations possible for these *Homo sapiens,* a theory of scale will leach into all aspects of how we encounter the world.

But, of course, it already has. The theory presented here is thus a means of examining and dwelling with this implicit theory of scale, untangling the contradictions born from the residue of nonscalar views, and thereby reorienting ourselves to the strangeness that is our modern technological world, a world that insistently and persistently reveals ourselves beyond ourselves and demands that we comprehend this dissolution.

PART I

Algorithms for a Theory of Scale

Scale is not the exclusive purview of science or the technologies that produce any scalar observations. Rather, these modes of inquiry and extensions of observation (e.g., in microscopes, telescopes, and the many more complicated apparatuses) examine basic aspects of experience that already generate the basis for scalar observations or necessitate the introduction of scale for the understanding and comparing of empirical observations.

On this premise, the following algorithms are fractal unfoldings of principles and definitions that are derived first through three thought experiments that I call "experiential origins of scale." By this I mean basic phenomena that arise in experience that create the need for scale and the basis for producing it. These experiential origins are not produced by first creating an apparatus (as is the case with the creation of a map) or by attempts to understand pre-given objects, but rather through changes in one's approach to observation. These observations are then augmented and extended using technologies that permit us to comprehend this extension beyond typical human experience. This approach positions scale as an attempt to understand basic questions about experience and observation.

I then add three additional "thought experiments *in* scale" that assume the notion of scale laid out in the experiential origins. These thought experiments are more properly experiments in thinking in terms of scale, and they are meant to train the reader in scalar thinking. Along the way I propose some more speculative reconfigurings of some scientific assumptions and approaches. Thus, while the experiential origins are meant to follow a purposefully naive approach to inquiry and observation, the experiments in scale are meant to pave the way toward a more sophisticated and far-reaching extension of the power of scalar thinking.

To show the continuity and application of these points, throughout this work I will refer back to the points made here using the paragraph numbers.

1

Distance and Resolution

The First Experiential Origin of Scale

1.1—Let us start with a common observation. In front of me is an object: a tree. If I move away from the tree, the tree gets smaller. If I were to keep moving away from the tree while still keeping it in view—going to the top of a mountain, perhaps—I would find that the tree loses more and more detail until it merges into another object: the tree becomes part of a new entity called forest. The phrase "you can't see the forest for the trees" becomes more meaningful as a statement about how the entity "forest" is a larger phenomenon that is not perceptible from the same position that one sees "trees." The basic experience is that, in the view of trees, one does not discern this larger object, forest. This truism points to how this observation is not in any way new. Yet, if considered in its simplicity, it can unfold into a series of more significant observations about observation that generates the need for scale.

1.2—Clearly there is something fundamental about the brain's processing of a visual field that produces this shift. However, if we were to proceed through the science of optics, searching for aspects of the processing of visual signals that produces these shifts in appearance, we would only find a mechanism for producing this change.[1] But tracing such a mechanism does not change the fact that this visual shift occurs or explain its implications, but rather tempts us to dismiss it as an artifact of vision rather than a general phenomenon about all observation. (What follows applies to all perceptions, which must contain similar mechanisms for separating out thresholds of perceptibility according to the scale of relevance. Vision, however, makes the shift in scale easy to see [as the metaphor implies] and track with sufficient precision.) We must pause on this simple observation: the change in the relative position of observer and observed leads to an alteration in what is able to be distinguished as an object.

1.3—The view from the mountain already suggests a fundamental characteristic of this shift: that individual entities dissolve into different entities when we change the scale of our view. What on one scale appears separate, on a larger scale appears to be one. This is not just a set of entities considered together—a set of trees. Rather, we find a whole new entity—the forest. If we could imagine forgetting that trees exist and only consider these phenomena via the perception of forests, we might consider the nature of "forests" quite differently.

1.4—If we pause here on the mountain, we find it hard to forget, however, that these forests are made of trees. There are at least four reasons for this. First, we look around us and see trees close by and relate them to the forest in the distance. We depend on what is near to provide a perceptual premise to understand more distant perceptions. This amounts to a reassertion of the scale of proximity as a mode of interpreting distant objects on familiar terms.

1.5—Second, our memory and language encode these transformations of perspective in a way that often obscures the shift. For example, "grass" already implies an aggregate in the same way that "forest" implies an aggregate of trees (and other objects). Walt Whitman's famous "leaves of grass" is already a play on this kind of scalar encoding. Thus, in "A Song of Myself" he moves from an opening scalar identification ("I CELEBRATE myself, and sing myself / And what I assume you shall assume, / For every atom belonging to me as good belongs to you") to the image of grass ("I loafe and invite my soul, / I lean and loafe at my ease observing a spear of summer grass"). The awkwardness of singling out a "spear of grass" complements the opening scalar observation by highlighting the ease with which we speak and categorize certain things in aggregate and certain things in part.

1.6—Third, the object "forest" retains an intimate connection with the object "trees," such that, in viewing the forest, we don't think that we've lost sight of trees. Rather, we can retain the sense that in seeing the forest we are seeing many trees together, because the shift is not so great that the object "tree" is rendered completely unfamiliar. Indeed, if a tree is standing by itself, it might be perceived as a tree even from the mountain.

1.7—Finally, the consistency of the change in observation makes the shift seem trivial. The change in the appearance of objects is easily dismissed as a result of distance when we observe that moving away from an object creates this transformation in appearance.

1.8—If we remain at the mountain, our notion of scale will simply be the relation between these relatively familiar objects: the trees here and the forest in the distance. We arrive at what I am isolating here as a *scalar perspective* when we move even farther away from the objects at hand. If we continue to move away—perhaps now via a camera on a spaceship—we will find that even the forest disappears as a distinct entity to be discerned. The key is that, as you move farther away from an object, the familiar objects are entirely transformed, such that, if we were to experience wholly at this new perspective, the familiar objects (both tree and forest) would not be distinguishable at all. At the scale of the planet, as forests give way to colors on a landscape, trees are no longer observable in any way that could clearly be tied to the current observation except via a scalar shift (in this case, literally coming back to Earth).

The significance of this point is clearer if we, somewhat prematurely, go to smaller scales. The discovery of cells, microbes, atoms, and subatomic particles required the delineation of wholly new objects made available for observation at these different scales.

1.9—To mark this complete transformation of objects, we must introduce the notion of "threshold." A threshold is a discontinuity in what is observed, a change in the kind and quality of information that can be perceived as one alters the scale of observation. The need for scale arises when you observe this threshold shift whereby the whole experiential field transforms into new objects and need to make sense of what has happened. (We will arrive at these thresholds again in a different manner in 3.27 and continue to refine the notion of threshold throughout these experiments.)

1.10—Notably, this need for scale arises *before* measurement (the starting point for the second experiential origin); it is not measurement alone that produces scale, and scale is not itself simply a measuring process. Rather, scale arises first from an observational shift in the objects able to be perceived. Thus, scale relates first to observational thresholds rather than physical thresholds, since the first describes changes in what might be discerned while the second describes changes in the scale of an object (an object getting larger or smaller), which presumes that object in advance. While we might characterize this threshold as a phase transition in perception, we can reserve the term "phase transitions" for a later point (see 6.19).

1.11—Considering this observational shift in terms of distance highlights the continuity of the perceptual shift from one scale to another across any scalar threshold. Scale requires this continuity, since this permits us to trace the change in observation across this threshold. We will arrive at other reasons and implications for this continuity later (2.6, 6.6).

1.12—The continuity in the observing apparatus is met equally by the discontinuity in what is observed. This point cannot be emphasized too much: scale entails a consistent and accountable alteration of observation that produces radical changes in what appears in any observation.

1.13—Prior to the 1960s, continuing our experiment in distance via literal movement of the eye was not possible. This points to the essential relation between technology and scale: scale requires technological adjuncts, because scale is only produced when we move beyond the perceptual limits of the human body. Thus, we can rewrite the previous three points: the significance of scale emerges when the lifeworld of objects available to the *Homo sapiens* perceptual apparatus encounters a perception that exceeds these limits.

1.14—Even in going to space, you haven't physically switched scales. That is, the body of the *Homo sapiens* has not become as large as the Earth as if one has eaten the appropriate cake in *Alice in Wonderland*. Thus, you still have the same problems we noted above (1.4–7) at the top of the mountain: you can look around and see objects at the meter scale. However, once we have crossed a scalar threshold, one must ask: what is the relation between these two perspectives? Then one arrives at scale in the disorientation between two radically different perspectives. This is essential: in scaling, we are dealing with two very different experiences and putting them in relation. Scale is not either of these experiences alone but is an experience about two experiences (see 2.5).

1.15—Here, scale results from converting this observational movement into the concept of resolution. Resolution is the amount of detail one can discern within an observation. In relation to scale, resolution can be defined as the range of objects that are able to be taken as information for differentiation at a given observational range.[2] At different resolutions, different objects are discerned, so that one sees not merely small parts of a material but whole new objects. Shifts in resolution and shifts in scale go hand in hand: scale tracks the range of observation, while resolution points to the amount of detail able to be seen at that range.

Converting distance to resolution is truer to how scale moves beyond the human body (1.13). It also makes it possible to scale to the smaller, since simply moving physically closer to an object will not produce a revision of the world of objects unless accompanied by a change in resolution.

1.16—*Scale domains* can be defined as the ranges between thresholds of observation, where the field of objects revises into an entirely new set of objects. Within a scale domain, objects remain relatively recognizable, as the tree does even from the mountain. Thus, a forest remains within the same scale domain as the tree. It is only when we move farther, and the forest becomes a landscape, that we find ourselves shifting scale domains. In the remainder of this project, the phrase "change scales" indicates a shift across a scalar threshold into a new scale domain.

1.17—To change resolution is to change scales even if you don't move. Thus, if we were to remain in orbit and change our perceptual apparatus so that we resolve a singular tree, then we would be changing scales (scaling our perception via change in resolution). This meets the objection that better vision would allow one to view two scales at once or that certain parameters of the apparatus (e.g., the eyesight of a hawk who is able to see a mouse from the sky) cross thresholds of intelligibility.

1.18—The relationship between scale and resolution helps us understand the great innovation of the microscope. The microscope brings out an essential question of a threshold: do we conceptualize these objects in the terms of normal scale experience? We have arrived at a scalar perspective when we no longer consider the things in the microscope as simply really small objects within the same scale domain (cells as small compartments in the skin) but rather as a whole new set of objects (cells as units of their own).

1.19—In this scalar view, we can define objects as those differences that are able to make a difference at a particular range of observation.[3] By "differences" here we mean the grounds for separating objects available within a particular resolution. The tie between differentiation and resolution means that, to some extent, differentiation depends on the level of observation (i.e., scale). We should thus distinguish differentiation from division, since division implies an action taken on a pre-given object rather than a shift in observation. While any given scale provides grounds for differentiation, one must first select the resolution at which differentiation is to be produced.

1.20—Taking resolution (1.15) and the consistency of observation (1.11) together, we can state that scale produces multiple ways of looking at the same "thing" such that it will appear as different objects depending on the scale of observation.

1.21—The major premise implied by this shift in resolution is that any given differentiation might be rendered otherwise. Returning to 1.3, we can state the conclusion more generally: what appears differentiated on one scale is differentiated differently on another scale. That is, both differentiations and objects depend on the scale of observation. This implies that objects and differentiations are tentative or secondary delineations that depend first on the selection of the range of observation. At the same time, this does not mean that the observer creates the differences available for differentiation at any scale, but only that one must first designate the scale of relevance.

1.22—The unity of seemingly separated objects becomes available as a shift in perspective, not a bringing together of previously ontologically separated entities. As one moves to a larger scale, objects become a single unit as disparate things come together into a new object able to be discerned at that scale.

1.23—These units, however, are in turn inevitably parts of larger units. Scale to the largest possible scale imaginable. We can say that the "thing" that is being scaled is itself already "what is," or the Cosmos. By definition, one could not observe this "thing," since one is inside it and part of it. However, only such a unit that includes all units could be considered a "Whole." As a resolution, this Whole would be observationally One, since no second object could be distinguished apart from it.[4]

1.24—Multiplicity is the discovery of differences that one might resolve within a previously unified object.

1.25—Unity does not annihilate multiplicity but includes it within it as a function of a change in resolution (see also 3.39).

1.26—Scaling smaller reveals multiplicity (more units within a unit), while scaling larger reveals unity (a unit that encompasses given units). However, this logic breaks down at the scalar extremes: there is no reason to assume that reality must continue to produce discernible modes of differentiation at smaller and smaller resolutions. Likewise, the scale of the Whole does not produce a unity in the sense of an object able to be seen together. Thus, both ends of the scale are equivalent to substance in Spinoza's sense, a term already indicating that which underlies, that is, that out of which differentiations might be discovered to discern objects, and equivalent to the Bhagavad Gita's concept of *kūṭastham*, the unchanging, or "that which stands at the top or the highest position."[5]

1.27—The condition of possibility for redifferentiating this same "object" (whether the Whole or any object within that Whole) is that one might divide the world differently. This necessitates a kind of momentary and conceptual zeroing out of the mode of differentiating objects so that one might resolve them differently (see also 2.9). Thus, scale points to the inessentiality of objects and the unity of reality in three different senses via the Whole, the substance, and the negation of (which equals the openness of) differentiation.[6]

1.28—Only if we remain on one scale can we assume that an object discernible at that scale is contained and separate in itself. To adequately capture the paradoxical nature of this persistent habit, we can name this *the partial-whole problem*. The partial-whole problem includes all forms of distinctiveness taken as absolute or grounds for the distinction of objects as separate in and of themselves—in short, as "entities." This includes any kind of haecceity (Don Scotus), monads (Leibniz), holons

(Koestler), assemblages (Deleuze and Guattari), autopoietic systems (Maturana and Varela), and individuations (Simondon) that might be distinguished *insofar as they are considered distinct in and of themselves.*[7]

1.29—Because scale crosses us into thresholds beyond our human bodies, we need to be on guard for the ways we apply nonscalar experience to this scalar experience. Nonscalar experience leads us to persist in assuming the pre-given nature of the separation between objects able to be discerned at any given scale. It also tempts us to remain on one scale rather than being open to other possible scales.

1.30—The privileging of any given scale we can call *scalism.*[8] The first scalism is always going to be a privileging of this scale (around one meter), since it is the scale at which *Homo sapiens* live. This meter-scale reality is the nonscalar experience out of which our whole lifeworld, sense of reality, language, and culture is built. However, we can see how other scalisms are easily put into practice, for example, in strong reductionism's assumption that a smaller scale holds more explanatory power than a larger one or the reverse assumption that larger scales inherently control smaller scales.

1.31—This experiment permits us to distinguish what we are here calling "scale" from scaling in the sense of making objects bigger or smaller. I will call this *Gulliver's scaling*. Gulliver's scaling presupposes objects, keeps them intact, and considers making them larger or smaller. Thus, in *Gulliver's Travels* we find very small and very large versions of essentially the same lifeworld of human beings. We find similar operations in *Alice in Wonderland*, Godzilla, King Kong, and similar scalar distortions (more on the nature of these in 2.19). However, this notion of scaling also operates in biology in the idea that organisms have limits to size and scalar proportions in growth (see 3.30).[9] Scaling operations, in the business sense of making an operation function in a wider range of effect, are also a form of Gulliver's scaling.[10]

1.32—We can also separate out cartographic scaling. Cartographic scaling is about representation, while scale here has to do with specifying shifts in observation. As we will discuss later (2.15; chapter 11), these two are related, but to treat this larger apparatus of scale as the same as cartographic scaling is to obscure this essential relation between the transformation of observation and the transformation of reality.

1.33—I am separating out both Gulliver's scaling and cartographic scaling from what I am defining here as "scale itself." I'd argue that the notion of scale I provide here is more fundamental (and thus deserves the title "scale itself") because it extends beyond these changes in objects and underlies these representations. Scale has its own rules, conditions, and possibilities that are derived apart from any Gulliver's scaling and representations.

2

Measurement and Perspective

The Second Experiential Origin of Scale

2.1—Our trip up the mountain in chapter 1 suggests another experiential origin of scale. Given that this change of perspective generates two very different fields of objects, a need arises to compare and relate these two perspectives together. From the perspective of the mountain, how do I find a reference point for understanding the objects that are now in view? The answer: a measure. But how does this measure work, and why is it needed?

2.2—As we noted in our move up the mountain, my usual sense of size is formed by the physical proximity of the body to the structure called "house" (for the sake of quotidian variety, we'll switch from observing trees to houses). If I stand in front of the house, it can take up most, or even all, of the visual field. And yet, in moving away from it, that house takes up less and less of my view. From the top of the mountain, it is hardly noticeable. Our first experiential origin kept us moving further and permitted this horizon of difference whereby "house" could be distinguished to disappear. What if, instead, we take a more rational—literally, ratio-based—approach to this experience and provide a reference for the phenomenon's transformation?

2.3—Thus, we return back to the house and decide to insert a reference to help determine this thing we call "size." The selection of this reference is almost arbitrary—we could choose my hand, a car, a random stick, whatever—except that it will be more convenient if we can take it with us, hold it up to things, and, most importantly, see the object in our immediate experience. Thus, using the whole building as the measure would not be particularly useful, nor would choosing a speck of dust. Truly scalar objects are likewise useless—you can't use galaxies or cells as your base measure in this experiment. Likewise, the reference itself cannot change in length as we use it—so it needs to be a relatively stable object. Importantly, once the reference is chosen, *it is no longer arbitrary*. Now it has become a measure. For the sake of familiarity, let's use a meterstick.

2.4—If we want to measure the length of the house, we hold the meterstick up to the house and count the number of times it goes across the house. If we want, however, to then measure the house as it appears from the mountain, we have a problem. We can hold the meterstick up in front of the eye and find that it is useless for measuring the house. Since we can't put it up against the tactile boundaries of the house while this far away, this measure is completely reliant on how close the meterstick is to our eye. Suddenly the measure itself seems oddly variable despite being chosen for its consistency: if I bring the meterstick close to my eye, it appears that the whole landscape measures one meter and the house perhaps a centimeter, but if I pull it away from my eye, suddenly the house will measure several centimeters. Of course this is absurd, because it is not how we produce a measurement. What has happened? The distance from the house has produced a change in the way the house has appeared. We can thus approach the reference another way, measuring not the object as it appears for us, but *the eye's distance from it*.

2.5—Measuring the eye's distance from the object will produce a reference point for a ratio between the way the object appears and the size of that object depending on the distance from the object.[1] This is, in fact, a somewhat simple mathematical maneuver that is essential to the work of astronomy.[2] What we have done in this operation is move back and forth between the two positions as a function of distance in order to determine the comparative size as it will appear in a visual field. Here it becomes clearer that scale is created only by the relationship between these two very different perspectives (1.14).

2.6—What has changed in the scaling operation? Not the meterstick, the perceptual apparatus, nor the object. Rather, the *position* of the eye has changed. In other words, the *perspectival shift* creates the change (see 1.11).

2.7—What, then, is the measure measuring? As a function of distance, it is measuring the position of the viewing apparatus relative to the object (the eye to the house). But as a function of ratio, we are measuring the relationship between one view (from the mountain) and another view (from in front of the house). In scale, one is *not measuring objects, but perspectives.*

2.8—Furthermore, one of these experiences or perspectives will always be inaccessible or atypical in some way. From the mountain, the house is "over there," and the view of it is different from the usual way of interacting with it. If we went back down to the house, the experience of viewing it from the mountain would now be "over there." In each case, the "here" remains the same, based on the immediate domain of proximal objects, which is compared to the atypical perspective. In fact, the measure was chosen precisely for its ability to refer to this normalized, proximal experience while serving as a reference point for the distant object. The measure provides a consistency based on typical, local experience to make sense of atypical or inaccessible experiences.

2.9—The presence of the objects measured can lead us to forget that we are also measuring perspectives. But the role of perspective is further obscured by that fact that in measuring we attempt as much as possible to treat the *position* of the perspective as a zero point. In "putting the meter against the house" we attempt to zero out perspective by removing any intervening angle or perspectival alteration between the measure and the object. At a distance, we cannot zero out in this way. Instead, to measure properly, you have to take the distance of the eye into account, treating the eye as the place from which the measuring starts but which itself is not added to the measurement. Oddly, this zeroing out *is* to take into account the perspective.[3]

2.10—With a measuring standard in place and a sense of proportion, we can produce scale from the measure itself. Since the measure already contains a reference to your usual experience, it can function to relate *any degree of perception (even nonvisual) to our usual experience by applying the proportion to the measure itself rather than to the visual difference.*

2.11—Here, it becomes clear that scale is not a percept but a concept: scale uses a measure to provide a conceptual relation between two perceptions. Scale is a marker about perception, providing a reference for accounting for variations within perception. But, as a concept, it can also be extended far beyond this proximal experience that we used to produce the relation. In doing so, the concept has not lost its relation to perception but rather encodes this relation—including a standard of consistency and a tie to empirical shifts in observation.

2.12—The measure can now be used to account for observations beyond our experience by increasing the number of measures or cutting up the measure. This compounding and cutting up is the projection of a relation to normal experience that permits us to retain the essential scalar ratio. Doing so produces scale as we are familiar with in science: it permits us not only to consider the house but the house on the scale of the nanometer.

2.13—What has changed in this cutting of the measure? With the size of the nanometer, we have changed the size range of our perspective while retaining a reference to normal perspective. The reference "nanometer" becomes a signifier of the scale of observation.

2.14—These *Homo sapiens* bodies will never really perceive the nanometer scale in the same way they perceive the house. Yet the reference "meter" preserves the relationship between that scale of observation and the scale of experience, at which a meter is clearly discernible. From this perspective, the great innovation of the metric system is that it retains this reference to normal experience clearly in view by simply compounding a measure selected from the scale of normal experience. Such measures embed the appropriate degree of consistency (2.6, 1.11) into the scalar shift.

2.15—In this extension or dissection of the measure, this second perspective—the "over there"—has now become embedded so that, insofar as we try to take it on as an actual perspective, it becomes mediated and represented. When we find ourselves with representations of objects on a scale exceeding the human lifeworld, the whole process becomes further disconnected from perspective. *Now, looking at the map or the picture, we are tempted to state scale as a relation between two objects (the image and the thing being represented)—and this is incorrect.* Rather, when we are looking at the object that contains a scalar image, we are not experiencing that image as an object but rather as a perspective.

These perspectives have been navigated by the scalar relation and embedded within the representation.

2.16—In representation, scale retains the measure as the statement of a relationship to normal experience. The ratio on the map or image says not just "this object is *x* times smaller than the landscape" but "the view you are having of this object bears *x* relationship to your usual experience of the landscape." It is the latter observation that permits us to use a map to maneuver through a landscape. However, this is again why we must distinguish cartographic scaling (1.32) from scale here: representations inevitably have other alterations, filters, or distortions that are a product of the representation rather than the shift in observational scale.

2.17—We can now understand how scale uses a measure to orient us to technologically produced representations. Scale says: "This picture you are viewing of Jupiter would be the equivalent of 139,822,000 of these meters you are familiar with" (Figure 1). I cannot perceive that number of meters stacked next to each other except via this scalar compression. Projecting out the measure taken from our usual perceptual field makes the incomprehensible comprehensible, but in a limited way by using our reference to compare this phenomenon (Jupiter) in relation to our usual phenomena around us. In sum, the scalar notation is a sign that allows the perceiver to perceive her own perception (here, the picture of Jupiter) as it is projected out beyond its usual perceptual constraints but, through the scalar consistency of the measure, rendered in ratio to the phenomenon around them. Without a scalar reference, someone who had never seen Jupiter would have no reason to think it was an object much larger than our whole planet (this could, in fact, be a beautiful little marble or a sphere suspended in a liquid).

2.18—If scale is the relation between one "over there" and another "over here" (2.8)—or the equivalent in the more projected forms of observation—then where is the perspective that is scale? Properly speaking, a scalar perspective is in both places at once: *it is a situated dislocation.*[4] The microscope or telescope does not avoid this problem, but rather more properly takes that "over there" and transforms it, that is, scales it, so that we do not even need to come down from the mountain to see the house as we would standing in front of it. In viewing the house with a telescope or, in a further abstraction, just viewing the image of a

FIGURE 1: This is how to see 139,822,000 meters. Such images not only present an object but represent two perspectives, as long as the scalar relation is in place. Image by NASA, JPL, and University of Arizona.

cell or a planet, we may seem to have made these two perspectives into one. But the making-one of the perspectives is only rendered scalar if within it is buried the reference to two.

2.19—Consider, in contrast, an example of Gulliver's scaling (1.31): the distortions in scale that can occur in film. Prior to computer-generated special effects, filmmakers used the relationship between vision and distance in order to distort the scale of objects.[5] For example, to produce the sense of enlargement, two shots taken at different distances can be overlaid onto each other to produce the sense that something is far larger than it is. Likewise, one might use a proportionally smaller model, as in King Kong. In the former case, two different perspectives are presented as one, which voids scale by overlaying these perspectives without an intervening scalar measure. In the second case, the scale model is presented *as if* it is a normal-scale perception such that the distortion is not taken into account. In both cases, measurement and ratio are impossible. In fact, the "effect" occurs by obscuring them. If we treat the image in the microscope or telescope as functioning in the same way, we miss the fundamental difference in perspective that creates the phenomenon of scale (we will return to this problem in chapter 11).

2.20—*Scale holds oneself to a measure.* This is not to say that the measure is being imposed onto oneself in some arbitrary way. Rather, the consistency of that measure gives one a consistent reference point for tracking the transformation of objects as a function of a transformation in perspective. This "transformation of perspective" is not a transformation of something external to the whole operation out of which the phenomenal world appears. Therefore, the measure is one necessary means of holding oneself consistent for changes in any observation.

2.21—Scaling in the microscope or telescope can be seen as systematic, consistent, and measurable distortions of the apparatus. Likewise, alternative modes of "seeing"—for example, in the atomic-force microscope, protein crystallography, neutrino detection, and so forth—are coherent perspectives as long as they are put in relation to some measure derived from normal experience.

2.22—But in transforming this perspective, the question remains: what has been transformed? What is this entity who is perceiving such that it is capable of such a radical transformation? Our first inclination is to say "the apparatus." But, crucially, even the microscope or telescope *does not change*. Rather, the device is added to perspective so that the eye sees differently. Perhaps we could say that the compound apparatus eye-microscope has changed, but even so, the consistency that renders it all measurable—and therefore accountable as a scaled perception rather than a distortion—requires that all of these factors be taken into account (e.g., accounting for the magnification power being used) in a way that also includes the object, the properties of light, the effects of gravitation, and so on. In less visual or direct modes of observation (e.g., protein crystallography, colliders, neutrino detectors) the problem is only more marked. What and where is that "thing" that sees (or detects or pieces together differences) if all of these are part of the seeing?

2.23—The final step is thus to turn the same apparatus on oneself, not just in the measuring of perception but in the taking into account the one who can maintain and compare these perceptions across these shifts such that one can have, at once, two very different perspectives. The real difficulty will always be, in the situated dislocation (2.18), where is the one who has this "perspective"? Who is scaling? In scaling is there any object, for example, the eye, that can be said to possess the perspective? But, as we have seen, the eye that scales is always already in at least two places at once (2.8), must be taken into account (measured) (2.7), and must be treated as the zero point from which the measuring begins (2.9). But, since scale produces an experience, this question of pinpointing the perspective makes more sense by referring not to each of these individual points of perspective but by reference to the experience of that dual location (1.14, 2.5), and the transformation between them. This dual location is the first way that scale dislocates the subject—subject taken here in the formal sense of the one for whom an object appears.

2.24—This dual location introduces a tension between the production of perspective and identification of the "subject" who has that experience. If we now start from a scalar perspective and ask "who scales?" we can see a further dislocation of the notion of subject, subject now in the sense of identification with an "I" who experiences. Our usual sense of identification might be rewritten as a scalar error: within the normal scale of experience, an object has been selected to be the subject—that to which experience is ascribed. But if we switch scales, acknowledge the tentativeness of objects (1.21), and acknowledge that scale places that perspective in two places at once (2.8), then we cannot likewise ascribe experience to a particular, clearly distinguished object definable in and of itself on one of these particular scales. In other words, we need to extend the partial-whole problem (1.28) to the sense of identification. We can call this the *ego-structure problem* and treat it as a subform of the partial-whole problem.

2.25—Just as we suggested in 1.27 that scale required a momentary zeroing out of differentiation, we must likewise suspend absolute identification with the scale of experience in order to properly grasp a scalar perspective. In other words, "I" cannot identify with one or the other of the perceptions bridged by a scalar reference if the experience is going to be truly scalar.

2.26—The transformation here is thus also a transformation of what one holds to, the experience held to be normal, and the assumption that one will be inevitably emplaced within any apparatus taken as an object on a given scale. Most importantly, scale transforms our sense of our "selves" as the *Homo sapiens* body only definable at the scale of experience. In this sense, the ego-structure problem preempts or perhaps holds in place the partial-whole problem, since the egoic identification provides the grounds for scalism (1.30). It is one thing to attempt to zero out the divisions of the world and another to zero out one's whole complex identification and structure of values built around this-scale experience.

2.27—If one no longer looks for any particular object on any given scale to identify as the subject, then the apparatus or "one who scales" is thus, in a literal sense, the Cosmos itself. That is, if one cannot delimit within the Whole any object to *have* the experience, then the experience is, properly speaking, open to these possibilities and comparisons only because there is not a coherent, separate, divided subject experiencing these things from the outside. Instead, the Whole itself is that out of which, in which, and for which such divisions are made, preserved, compared, and extended.

2.28—The particularity of any experience could then be defined as a specific configuration of the Cosmos delimited within itself in a particular way. If so, properly speaking, within any view on any scale is already contained the whole of what is, filtered according to the particular parameters of that experience. However, for reasons that will become apparent later (5.11), a majority of the universe is too minuscule or too extensive to be registered directly within an experience in a way that produces the thresholds being registered with scale.

2.29—We have, then, projected our "selves" beyond the confines of the "I" and rediscovered a sense in which, properly speaking, this "I" is the Whole experiencing itself in the sense articulated in the Hermetic microcosm/macrocosm or Vedanta's Atman/Brahman. The primary function of these articulations is to transform perspective held to be "mine" for the sake of scrubbing ourselves of the scalism developed out of the habits of this-scale experience and values.

2.30—One might, via another apparatus within the Whole, trace out the significant structures, delimitations, and filters that produce any given scalar experience, using the thresholds provided by scale (1.16, 5.11) to identify a series of relations that produce one observational configuration. Thus, the perception of a protein molecule can be traced, as the anthropologist of science Natasha Myers has done, in a cross-scalar configuration of proteins, crystallographic techniques, polymerase chain reactions, microscopes, modeling, animated gestures, laboratory setups, funding streams, artistic renderings, personal motivations, dance competitions, and so on.[6] It is in this sense that scholars in science studies have emphasized the networks and apparatuses of science. However, this is a secondary tracing that already presumes another apparatus tracing the process of knowledge-making. We will take this point up in chapters 7, 8, and 10, since even these notions of networks or apparatuses do not fully account either for the scalar nature of this tracing or for the way that they dislocate identification even as they situate an experience.

3

Scope and Accumulation

The Third Experiential Origin of Scale

3.1—A final experiential origin of scale can be found in quite a different approach: in terms of the accumulation of knowledge. While we could begin with science, the social lives of humans present us with a starting point that begins from the simple inquiry into one's surroundings.

3.2—Looking around her and attempting to get better at this thing called "surviving," the *Homo sapiens* begins to inquire into the organization of her surroundings. She discovers or is told about some relations: this plant helps with this, these animals are good for that, these people are related to you in that way, and these all bear such and such relationship to ensuring that you get fed, don't get killed, and are able to reproduce. Inquiring into and remembering this map of relations is useful, even necessary for navigating the landscape of this *Homo sapiens*'s experience. Within this relatively immediate field of relations, most things are within reach and limited to what can be handled without much assistance.

3.3—In reality, the scope of consideration for *Homo sapiens* quickly extends into a larger organization of interaction and knowledge, which we can call the tribe, the estate, the city, or the like. Why? Beyond a single *Homo sapiens*'s range of experience, there is a frequency of interaction and an accumulation of data about relations that necessitates a move beyond any *Homo sapiens*'s limited experiential field and cognitive capacity. This creates a point of reference for a larger scope of consideration, relations, and organization beyond our immediate surroundings.

3.4—This higher point encodes a field of relations as a range of inclusion. The category—for example, the city—in itself is not an entity of consideration. Rather, the people, interactions, buildings, transactions, conditions, and so forth that are considered together as "city" are the topic of study. The term "city" indicates the interactions within a particular range, as the scope of the interactions to be considered together. To speak of the city is to speak of all of these interactions, and yet the city itself is not any one of those interactions. In this grouping, we find ourselves speaking of the city—its well-being, its structure, its status, its growth, and so forth.

3.5—At this point, the boundaries of the city need not be clearly defined. But the physical location of a city, as well as the limited scope that can be sensibly and usefully grouped within it, means that at some point the city itself becomes more clearly delimited. We can imagine this occurring in relation with another city. In this encounter, the definition and delimitation of the range of relations belonging to that city has to be clarified for the sake of a new set of relations that is now made possible: the relationship between one city and another city.

3.6—We have not quite arrived at scale, but now we encounter a form of logical typing. The theory of logical types was initially proposed by Bertrand Russell as a solution to a problem in set theory of the sort we are dealing with here. Set theory attempts to define the basis of mathematics using sets of (abstract or mathematical) objects. Logical typing becomes necessary when, as is inevitable, you begin to deal with sets of sets. A logical problem arises, known as Russell's Paradox. By way of example: a barber is a man in a city who shaves all those, and only those, men in the city who do not shave themselves. Who shaves the barber? If the barber shaves himself, then he doesn't shave himself. If he doesn't shave himself, then he does. The problem here is that we have two different levels of categorization or sets: the city, a collection of individuals; and the barber, one of those individuals. Because the city contains the individuals, the rule produces a contradiction, since it considers the barber together in one portion (as a city) and the barber apart (as an individual) in another. In response, Russell proposes the theory of logical types, in which you distinguish types according to the level at which you are considering. One arrives at a new level when you make a set of sets, for example, the city is a set of individuals. To fail to acknowledge these distinctions in logical types creates conceptual and interpretive problems, since you fail to acknowledge that what you are referring to is the same entity grouped in two different ways.

3.7—In encountering another city, a logical typing occurs: we become aware of the city as a distinct domain of relations to be discussed and put in contact with other entities that also exist on this higher logical type. But, even as we do so, we risk obscuring the fact that, at this higher logical type, the "city" is already formed and produced out of an aggregation of relations. Indeed, it is usually not so much the "city" that is making this recognition—at least not when considering the *Homo sapiens* within it—but a part of the city taking reference to this larger domain of relations.

3.8—The nuances of this logical typing are further confounded when, as more and more cities find themselves tightly woven in relations, we find ourselves defining nations as an additional logical type above cities. In turn, nations find themselves in relation with each other, and we start to speak of global relations. Each logical typing provides another name to describe a scope of relations. Each expansion of range does not erase the previous set of relations, but instead groups them according to levels at which we consider these relations together. But we must nonetheless attend to the level at which these relations are categorized and designated. To study the interactions within a city involves a different set of interactions that does not necessarily apply in the same way to the interactions between cities, and likewise in nations.

3.9—In this scheme, we have not found scale so much as layers of how humans might group together their relations. A logical typing does not immediately imply scale, particularly since one could create such sets even within the same scale domain according to any delimiting criterion. We find ourselves with the example of the city because of its tie to the basic *Homo sapiens* experience and social life, which provides a basis for the delimitation of logical types. When we apply this same structure not to social organization but to knowledge and inquiry more generally, then we arrive at a more interesting and thorough scheme that brings us to an account of scale.

3.10—Above, we noted that the limits of knowledge are one element prompting us to move to speaking of the city. Just as physical limitations prevent one *Homo sapiens* from building an entire city on her own, inquiry into the nature of reality, broadly conceived, will always fail to be exhaustive. One *Homo sapiens* alone cannot study every interaction and knowledge available even in her vicinity. A limit is inevitably encountered within thought, language, and perception: as one adds more and more to any inquiry, one inevitably reaches a limiting threshold at which there are simply too many objects to consider.

3.11—So what do we usually do? We get someone else to help or rely on the work of others. In doing so, however, the individual comprehension must encode the inquiry and accumulation provided by this cooperation. In scientific work, for example, you cannot repeat every experiment that your further experimentation requires. Others have done this, and others have checked those scholars, and as long as you spend enough time in that conversation to be comfortable with proceeding, you can proceed with them.

3.12—Here we see in knowledge something similar to what we saw in our first thought experiment about perception. Science must always find itself grouping things together, speaking of species, types, systems, and so on which exist on a different logical type than the individual encounters that make up these groups. Doing so makes possible some knowledge of—or at least a way of functionally describing—larger patterns of interaction. Doing so does not necessarily sacrifice the dynamics that produce this aggregate discernment. Instead, this maneuver creates two domains of inquiry: the inquiry into any specific encounter with an object of study, and the aggregation of those entities being studied together.

3.13—Science produces this aggregation effect systematically so that it uses an aggregation of empirical data to go beyond any given empirical moment. This maneuver permits science to expand the scope of empirical inquiry beyond what is immediately experienced. While the technologies of the microscope and the telescope are necessary for expanding the scope of what is *in* present experience, it is the printing press and the circulation of observations that permit the cataloging and accumulation of observation that could extend to large groups of organisms and places.

3.14—In this accumulation, the conception of a larger population or species starts to take on new meaning, as categories (e.g., species, populations, geological formations) that function, much like the city, as reference points for an aggregate of relations. With these larger entities, a shift in thought can occur: you can start to consider their relationship on a new logical type that extends beyond any delimitation according to types of objects in question. Rather than speak of a giraffe interacting with a tree, you speak of these species of giraffe interacting with the larger flora of the savanna or, in turn, the ecosystem as a whole scope of relations.

3.15—There is a difference between studying a bat, bats, and an ecosystem in which bats exist. In the first, you are dealing with a particular creature, here in front of you. In the second, you are dealing with all possible specimens grouped together in a category (species). In the move to the ecosystem, however, a more significant leap has been made: the delimiting criterion by which you only study one group of organisms (the category, bat) has been removed, and instead a whole structure of relations becomes the defining factor. An ecosystem becomes anything that can be designated as relevant within a broader field of relations.

3.16—We have now run into something similar to our designation "city": the ecosystem becomes a domain of relations that includes all relations within a particular spatial scope. However, this domain differs from the city, since the city retains two kinds of forced exclusivity which kept this formulation from being truly scalar: the city is set apart and defined by the humans who would mark out (in their maps and territorial battles) its boundaries, and, more importantly, the city always *uses the human as the defining factor* for cities. In other words, even if it is delimited spatially, the city is a grouping of human relations, while the ecosystem does not necessarily retain a distinction with regard to the relevant kinds of relations.

3.17—Bats do not provide the same logical typing of their relations as do humans. The ecologist may attempt to delimit the ecosystem using spatial criteria as we do with cities: everything within X boundary becomes relevant for study. However, this presents a difficulty, since spatial delimitations on this planet are tenuous and not contained. Scientific studies then must make additional choices about how to delimit their ranges of study. They might attempt to use some kind of boundary phenomena, some of which are less tenuous and porous (e.g., a lake or an organism's body) and some of which are more so (e.g., a river or an organ). This attempt to delimit produces two difficulties. First, any spatial designation within the Earth is already going to have many relations that move across that boundary. This will be the case unless you expand the scope enough that even the ecosystem starts to take on a new set of relations at the scope of the entire planet. The planet then becomes the point of relation for interplanetary interaction (e.g., the sun's energy, the moon's gravitational fields, an occasional extraterrestrial object large enough to affect this planet). Only at the level of planet is a meaningful boundary present for delimiting possibilities of an ecosystem as planetary relations (climate, atmosphere, orbit, gravity). Such planetary interactions are then more clearly relations with something we can call ecosystem, which is now a description of conditions for smaller-scale relations, for example, what is available for bats (see 6.30–33).

3.18—Second, even within that spatial designation on the Earth, there exists more to be included than we might originally discern, a fact made clear by the microscope and other technologies that reveal more to be included within the spatial range already designated. The microbiologist arrives and queries the ecologist: are you also including these microbes in the water as part of the ecosystem?[1]

3.19—In these two problems we encountered a scalar threshold (one in the large, the other in the small). These now prompt us to specify what relations become relevant within a particular range of size. At this point we arrive at scale if we convert something like the magnification power of a microscope into the notion of a range of observation that is not based on any limit or boundary within space (i.e., the borders of a city or lake) but in terms of the size at which one examines that space. Doing so creates this scalar formulation: *to inquire on a particular scale means to include every difference able to make a difference* (1.19) *within a particular size range.* All potential differences able to make a difference become relevant for the study of a particular scale, even if we do not immediately find and register them. This also responds to the objection that one might raise to the first point in 3.17: that plenty of things (even small things relevant for bats, such as sunlight) cross the threshold of the atmosphere and the planet. In turning to size range, we become less concerned about the absoluteness of that boundary and more about the range of relations able to be discerned at any given size. In doing so, the individual logical types formulated by any smaller accumulations come together into a far more significant logical typing that is a true scalar shift: all relations seen at X size become a designation of the scale of inquiry.

3.20—To summarize this progression: in scale we take a range of inquiry and group it together according to common attributes (e.g., all Bats, all Humans). Scale then takes this grouping and turns it into a shift in scope tied to physical location (everything in Q boundary). This physical location or delimitation is then converted into a designation of size (everything at X scale). Now that we have arrived at this scalar configuration, a few essential attributes and implications need to be highlighted.

3.21—Most startlingly, rather than being about a selected grouping, *scale is inherently inclusive*. In fact, it's fundamentally about domains of inclusion, because it is about what can potentially make a difference—that is, form a relation—at a given scale (3.19; 1.19). This has been signaled in 3.20 by the persistent inclusive signifiers "all" and "everything."

3.22—The implication of this inclusiveness is that many "shifts in scale" that we speak about habitually are not properly shifts in scale at all, since they are limited by additional criteria. When we speak of "scaling up" in economics or politics—even in our moving to the city and nation—we have performed only a partial shift in scope. Such a shift might function well enough when dealing with the local realm of actions, where human relations are already clearly embedded in a series of interactions with a local environment. But when we speak of global capital as a new scale of relations, we are already eliding the rest of the significant relations made visible at the scale of the Earth. One cannot speak of global economy or global politics without speaking also of ecology, since the shift in scale occurs in the extended range of relations, not in the mere aggregation of a set of entities already thought to be significant (humans). In short, scale implies that the global scale must include everything contained within the Earth.

3.23—While "all humans" may refer to some aspect of another scale, it is not itself this higher scale. Humans, insofar as they are a definable organism *Homo sapiens*, exist on the scale on which these bodies exist. From a global scale, these humans are a part of a larger domain of relations. The conceptual limitation of our scope of consideration to the human has conflicted directly with the actual change in scale to the planetary that has occurred in practice, perpetuating the neglect that produces the cross-scale disruption of mass extinction, global warming, and widespread pollution. In short, ecology describes the scale of planetary relations, not economics, politics, or culture. Human relations need to be subsumed into ecological relations (as another aspect of ecological relations) alongside the aggregate relations of forests, animal populations, microbial accumulations, fungal networks, and so on.

3.24—In practice, scientists delimit their inquiries to isolate particular factors for study. The filtering of the ecosystem according to bats provides a guiding delimitation in the same way the *Homo sapiens* did above for the city; the "ecosystem" can become anything that is relevant to bats. In doing so, we need to acknowledge that this widening of scope has already admitted into the study the excess already potentially relevant at this scale. This is not, as it might seem initially, that much of a problem, since we can invert the formulation according to this delimiting criterion: if I study a bat population, I cannot study everything on the planet, but I can study anything within the planet as they exist for bats. In addition, science is not usually interested in any single bat but in all bats, and bats as they relate and live in portions of an ecosystem. The conclusion here is just the significant but difficult admission that more is present than can be adequately described. This admission makes apparent the additional criteria as limiting factors and admits that these limits, although functional, will always be exceeded in practice.

3.25—We can now more adequately generalize our notion of threshold and rearticulate in a different way the notion of resolution. The final step of this thought experiment moved us from delimitations of numbers of objects to how we are looking at space more generally. We either expanded our range of inquiry until we found a significant boundary at the planet by which we could delimit the ecosystem (3.17), or we turned to the microbiologist to discover more relations embedded within an object (3.18). When we apply this to the ecosystem, we can say that the selection of the range of the British Isles is not a scalar shift, but a limiting oneself to a particular location. Scalar thresholds do not occur because you chose a particular nanometer segment of the world, but because the relations and objects being considered are those that become noteworthy within the scope of a nanometer.

3.26—To examine a particular scale is, at least in principle, to open the examination up to everything of note at that spatial range. This "everything of note" indicates that to study anything on a particular scale is to enter into a field of relations that immediately extends to all objects of a similar size. A scale presents a particular field of relations that exists within its own scale domain (1.16). This inclusion does not mean that everything is relevant in a given study (e.g., studying bats); indeed, other criteria no doubt enter for determining relevance. But scale has here isolated a range of potential relations that becomes relevant depending on the scale chosen.

3.27—We can now combine these observations with the concepts of thresholds (1.9) and resolution (1.15). These delineations of scale domains or resolutions are significant thresholds at which the field of relations becomes aggregated and new levels or relations are discerned. When scaling larger from our normal world of objects, the noteworthy objects remain relatively within range until we arrive at the level of the planet. If so, then the most sensible delimitation of range is between our normal set of objects and the global, between local weather patterns (it is raining here) and global weather patterns (climate), between local geological features (here is sandstone) and the whole Earth's geological patterns, between a local species (the hummingbird) and the larger ecosystem, which ultimately must mean the entire potential to interact as far as this organism can potentially range, that is, the whole planet.

3.28—At each threshold of relations (i.e., scale) there are new layers of complexity that are tied to the new ranges of differences that are able to make a difference (i.e., be considered relations). On a different scale, one finds a whole new range of possible variation to include within the study. Going to the size of the cell produces not just more to consider but a whole new domain of interactions discernible at that resolution. Indeed, the selection of a scale of inquiry greatly affects what you are going to examine in any inquiry. But again, this is not as great of a problem as it seems initially. Rather, studies can provide useful results by taking advantage of the thresholds at which differences become relevant (see 5.10–12).

3.29—These scalar thresholds are further confirmed by how life has compounded itself according to logical types of size. As J. B. S. Haldane noted as early as 1926, all organisms have a particular size in which they function well.[2] This leads to some significant thresholds at which organisms must not get bigger but must compound themselves by making aggregates of organisms rather than larger organisms. Single-celled organisms work with molecules but can only get to a certain size, at which point the coordination of single-celled organisms becomes necessary for more complex systems. At some point, these become sufficiently intertwined that a new organism is formed that consists of cells. The shift in size thus requires a shift in logical type, which also yields a different scale of perceptual being as this new logical type forms modes of dissecting the relations discernible at that scale (more on these points in chapters 8 and 9). Thus, the perceptual world of the *Homo sapiens* does not exist in the same field of reference that its cells do. The result is that, as long as our cells have what they need, they will do what they do according to a structure that is not necessarily known or fully determined by the larger structure. In turn, the organism next higher to the multicellular world we live in is not groups of humans acting in consort but the biosphere itself, which has always existed as a scale of interaction but which is still in the process, perhaps, of tying itself together as a cybernetically organized entity capable of reflecting on itself.[3]

3.30—What we have found is levels of relations that form size-bound, scalar logical types. Each of these levels creates another level of objects that function in a different way to the multiplicity of entities contained within that domain. Contradictions can emerge when you relate an entity on one scale to one on another scale, since the scope under which they function and appear distinct is different. That is, when you move across scalar logical types (scale domains), contradictions emerge because *you are talking about the same object from two perspectives* (1.20, 6.14). We can mark these as *scalar contradictions* or *scale errors*, to distinguish them from contradictions that emerge from within the same scale.

3.31—To invert 3.26 into a given object on a given scale, we might say that examining an object already admits everything within it on the smaller scales and potential referent to any larger-scale entity it is a part of. These relations, however, would be *scalar relations* outside of the field of relations immediately available for that object at that scale (see 6.29).

3.32—As we've already performed above (3.23), our language and concepts have a way of dealing with scalar relations in the form of speaking according to "all X" that establishes a relationship between two scales. We have to be careful here: "all humans" relates to "all trees" in a different way than I might relate to the tree in my front yard. This logical typing allows us to relate across scales even as it allows us to distinguish them. For instance, "all bacteria" bear a special relation to the planet's regulatory functions that, through their aggregate behaviors, adjust the larger makeup of this planet and provide the feedback mechanism under which the Gaia hypothesis makes sense. But to say that "a bacterium" does so is a scale error (3.30).

3.33—Such incongruity underlies our struggle to understand our own relationship to the planet. I do not exist as "all humans," even though I am included in that set. Yet this "all humans" is significantly altering, in a dangerous way, the terrain and makeup of the Earth's resources and atmosphere. Yet, I am also included in the set "planet." The traveling of this relationship between logical types is necessary for understanding one's relation to ecology. But to assume that one becomes or acts as "all humans"—and thereby can change the world—is not only to misplace oneself on a logical type that does not apply, but may potentially lead one to counterproductive behaviors that neglect the level on which this *Homo sapiens* exists and acts.

3.34—*One can only act on the scale at which one exists.* While we know this intuitively, we do not always articulate social change in this way: we become focused on the results made visible on a higher scale (e.g., we need to save the planet; we need to achieve X political change), such that we risk neglecting the sites from which these larger patterns emerge. In other words, we mistake the larger-scale patterns for personal actions. Even when such patterns can be manufactured or altered in some way, this alteration arises only by connecting into a larger system that mediates this intervention on a new scale.

3.35—We can now attempt to distinguish and name the scale domains according to these thresholds of perception and relations. As per 1.16, a scale domain is distinguished at the point where the objects on the previous scale are no longer discernible, that is, no longer able to be held within the scope of one's examination without creating a scalar designation. From this definition, we have a surprisingly small number of scale domains:

quantum
atomic/molecular (nanometers)
cellular/unicellular (micrometers)
bodily/normal (meters)
planetary/ecological
solar system (range of gravitational field of a sun)
galaxy
observable universe

We will consider these in more detail in 5.8.

3.36—It is interesting that we have some difficulty naming what I have designated here as "bodily/normal" only because this is the ground zero of scale for those *Homo sapiens* defining them. Most terms are inadequate, since they indicate too much content about that scale; for example, the "human scale" implies that it is ours. As a reference point in experience, this normal scale is the perceptual and interactive field on which the bodies of *Homo sapiens* directly act without intermediaries or compounding structures. It therefore also includes very small insects, animals, plants, and nonliving structures of the size we routinely interact with.

3.37—Note that I have left out a number of scale ranges that we might be tempted to include. These might be called *organ structures* with reference specifically to the organs of bodies and organelles of cells (see 5.10). These are the structures that life forms as intermediaries between different scales but which do not themselves consist of a significant enough shift in size to entirely change the set of relations. The reason for this should be clear when we consider that the scales distinguished by geographers and political scientists—local, city, regional/state, nation, global—are reduced in this scheme only to two scales: the normal and the global. This is because, as we stated before (3.16, 3.22), city, state, and nation all preserve their human-centered reference (see chapter 9). All organ structures are not significant shifts in scale for the same reason: they keep their tie to an organism and only become significant within that organism rather than changing the whole field of relations according to the size of their interaction. It may be that organ structures are the primary sites for scalar mediation between smaller-scale components (atoms for cells, cells for organisms) and larger systems across a scalar threshold (see 6.14–15). If so, then cities, forests, nations, lakes, fungal networks, and the like might be more properly considered as organ structures mediating the relationship between meter scale relations and planetary relations.

3.38—Following the argumentative form of our move to the Whole as object (1.23) and subject (2.27), we can ask: if scale functions according to thresholds of relations that are, in some way, within the same object, then what is this object that we might study in inquiring into the world? The term appropriate here is "the All." By "the All" we simply mean "include everything" that could possibly be, not just in terms of the largest size range but all possible resolutions within it. Here the Whole or the One (1.23) is redescribed according to a scope of inclusion rather than a function of unity. Thus, we can describe that-which-is as the One-Whole-All to emphasize these three aspects made clear by the scaling process.

3.39—The All is not a resolution for observation. Rather, it is a threshold of thresholds; it is the inclusion of everything that might be included within any threshold. As the set of all sets, the scale of the All is of a different character than any given scale. Rather than being subject to positive delimitation, the All includes all scales at once but is itself not any one scale. The notion of the All is thus of a different character than anything within it.[4]

3.40—This All-ness creates two reference points for inquiry: the everything and things within the everything. This is one source of the distinction between the one and the many, whether you are speaking of Parmenides's two realms, the Upanishadic Brahman and Maya, or the Hermetic microcosm-macrocosm (see chapter 7).

3.41—Even infinite terms like the All contain within them the possibility of adding more. While infinite terms may seem totalizing, they are inherently built on this openness. An infinite itself is not logically totalizing, as is demonstrated by David Hilbert's Infinite Hotel: if you posited a hotel with a countably infinite number of rooms, all of which are occupied, you can nonetheless always add one more guest to the hotel. When a new guest arrives, you just move each guest from her current room (n) to the next room (n+1) and you've opened up a new spot at the beginning.[5] Scale performs a similar maneuver by fleshing out the All both by adding more scales at which new relations might be discerned as well as leaving open the very limits, spatially and temporally, where relations might be said to exist.

3.42—The scalar descriptions produced by science have greatly expanded what can be considered a part of the All, but they have nonetheless left this openness intact. If anything, science has only reaffirmed that there is more within any given and apparent set of objects and relations as it takes objects that appear to be solid and easily defined (e.g., the body called "mine") and adds to them the relations of cells and atoms, extensive ecological entanglements, and the thermodynamics of suns and whole galaxies.

3.43—Scalar descriptions permit us to hide the openness and not-knowing already inherent in the basis from which these descriptions are drawn. That is, scale permits us to create maps of relations that intelligibly function within and across those fields of relations insofar as we attend to the scale at which we are describing. But within these descriptions and our knowledge of them still lies a deep and essential not-knowing for three reasons: the descriptions of this All remain indefinitely open to what has yet to be included; these descriptions map relations on a scale apart from the entities themselves; and any knowledge of this vast monolith of description will not and cannot be contained wholly by this point of reference called "you" (more on this in chapter 10).

3.44—These broader implications are entailed by the basis of scale itself as an apparatus of inquiry and way of understanding our own observations. Here, at the end of these experiential origins of scale, we can put together a simple definition that distills these considerations: scale is the systematic accounting for significant shifts in a measured range of observation.

4

To the Bottom

The First Thought Experiment in Scale

4.1—In order for any thing to be said to exist whatsoever, a differential must exist out of which a difference can be discerned. Every differential occurs on some scale at which a fluctuation or movement is able to make a difference. If one goes to a smaller scale, any given differential is no longer able to be used to register a difference.

4.2—Thus, if we posit that there might be a smallest scale at which differentials can be perceived and consider a scale smaller than this, then we might properly say, as Parmenides does, that the world is both One and without movement. This lowest scale remains hypothetical, and it might be that we could extend observational capacities to identify differentials at this smaller scale. However, then we merely move to an even smaller scale to find one at which no fluctuations are discernible. Thus, the Planck length[1] might be interpreted as the smallest scale at which differentials might be discerned, but it need not be this length for our experiment to proceed. We'll call this Scale 0.

4.3—If we then return to the scale at which fluctuations (or differentials or quanta) are first able to be discerned, we can track out how such differentials may produce objects at each scale. Let's call this base level Scale A.

4.4—We now need to highlight two essential concepts for the history of physics: a wave is a fluctuation in a continuous substance; and a particle is a discrete object with a discernible difference and boundary, a boundary that marks a concentrated location of affectability.

4.5—I have used the term *fluctuations* to indicate something particular about the particle–wave distinction, as it already causes many conceptual

FIGURE 2: An example of moiré phenomena. These two circles made of straight lines produce complex patterns and a sense of depth when overlapping. Moire Lines by SharkD is licensed under CC BY-SA 4.0.

problems in quantum physics (indeed, the difficulty with particle–wave duality may be a result of the following observations).[2] While one might think of waves in terms of common this-scale wave phenomena (e.g., on an ocean), we can conceive wave behavior here as the aggregation of effects—the accumulations of differences that make a difference within what, at Scale 0, is a continuous substance unable to be differentiated.

4.6—The hypothesis is this: particle attributes (or discreteness itself) are an artifact of viewing how fluctuations from a smaller scale aggregate so as to produce a difference that is able to make a difference on a larger scale.

4.7—Wave interaction, diffraction patterns, and related phenomena, such as moiré and beat phenomena (see Figure 2), demonstrate how fluctuations aggregate through interacting patterns in order to create periods of accentuated and negated interaction.[3] When two waves interact, they create patterns where some of the fluctuations cancel each other out and others combine to form bigger patterns. If it is true that such wave interference patterns are generalizable as a principle of aggregation, then we arrive at a fundamental scalar pattern.

4.8—The fluctuations at Scale A, interacting with each other, will create patterns of larger and smaller effects that compound certain fluctuations and negate others. If this compounding is at the edge of a scalar threshold (1.9–10, 1.16, 3.27–28) with Scale B it will create a distinct phenomenon: observing from Scale B, the compounded fluctuations at Scale A will make it appear as if an object has come into existence with the boundaries at the threshold at which those aggregate differences are now discernible at Scale B. In other words, from Scale B, only the aggregate effects that make a difference at that level will be observable (able to make a difference at that scale); everything else will be unobservable at Scale B. In this case, discreteness at Scale B is a result of the aggregating wave patterns at Scale A (Figure 3). Thus, an object—or discreteness and particle attributes—becomes a function of the aggregated effects across thresholds of perceptibility/affectability.

4.9—The movement of this object, now appearing discrete at Scale B, might be conceptualized as itself a wave moving along a string. The "object" might appear to retain essential consistency at Scale B while actually changing place in the fluctuating substance at Scale A.

4.10—We have continued to speak of waves to provide this point about motion of larger-scale objects (at Scale B), tie these effects to wave interference phenomena, and connect these aggregating effects to probability waves. However, this can be misleading, since wave phenomena are usually conceptualized as being on the same scale and do not necessarily result in a sense of apparent discreteness. The point here is that, at Scale B, the aggregating effects of Scale A will appear to be wholly discrete—an object clearly distinct and separable from its surroundings. In reality, the "movement" in 4.9 would be a movement in the aggregation of effects as they compound in size across a scalar threshold.

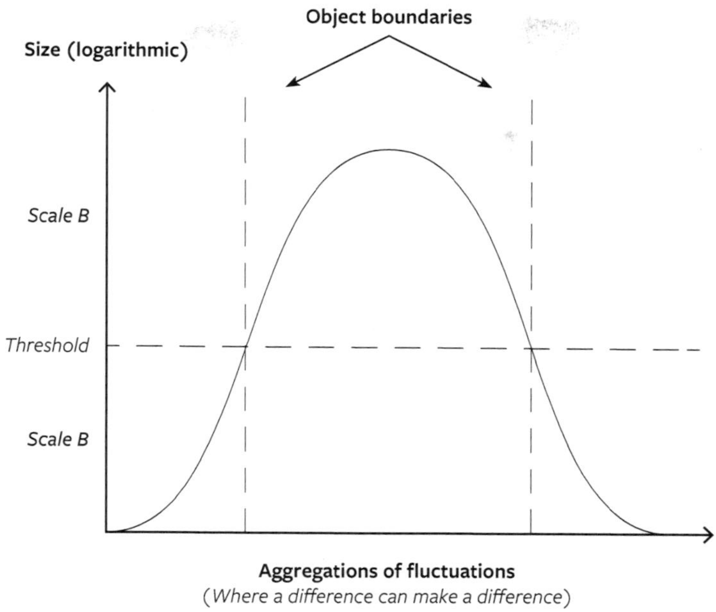

FIGURE 3: Objects as a result of aggregation. The curve is the aggregating of differences that are now able to make a difference at a larger size across a threshold between Scale A and Scale B. From Scale B it will appear that a clearly delimited object exists simply because aspects of the same substance do not aggregate to a sufficiently large degree to make any difference at Scale B, even though they still exist at Scale A. Image created by Graphit Science & Art, LLC.

4.11—If this thought experiment is valid, then the domain of possible objects at any given scale is determined by the pattern of aggregation on the scale below. Or, to put it another way, these objects are a description of principles or possible configurations of the aggregation of fluctuations on a lower scale. Thus, the possible subatomic particles are delimitations of possible aggregations of the scalar threshold below. In turn, the periodic table could be described as the possible aggregating behaviors of subatomic particles.[4]

4.12—After the initial move from Scale A to Scale B, the discrete attributes manifest at Scale B will now be the primary factor for determining what happens on the third scale, Scale C, across another scalar threshold. It seems that with each scale, discreteness plays a more important role, so that, for example, planets become far more significant than cosmic dust with less "stuff" that makes a difference in between. This, perhaps, is the significance that quanta originally played in the development of quantum theory: the quantum scale may be where discreteness emerges as a factor in determining the attributes of larger-scale phenomena. Then, on a larger scale, atoms stabilize this scalar aggregation and generate the possibilities for objects at micrometer scales, where molecular motion and forces are stable enough for patterns of energy to be harnessed in microbial organisms and cells. At the meter scale, volatility is further stabilized into the discernible and relatively stable objects of our lifeworld. Indeed, this provides an explanation for why Newtonian mechanics is able to operate alongside quantum mechanics: many have intuited that the issue is a matter of scale, but this thought experiment provides a potential explanation for why.[5]

4.13—Here discreteness also equals stabilization into relatively determinate patterns in comparison to the scale below. At each level, fluctuations seem to settle into set components with their own attributes and rules of patterning.

4.14—Such stabilization does not necessarily equal less complexity, as is well known in fractal patterns. However, to rediscover the complexity at a particular scale one can widen the scope of consideration, that is, consider a given scale near or at a larger scale. In comparison to the complicated coordination of objects inside a multicellular organism, the organism's body seems like a relatively stable and simple object. But if we consider that organism in relation to complex planetary interaction, complexity reemerges. Thus, we might say that *stability* is a description of viewing a given object in relation to its smaller-scale components, while *complexity* is a description of viewing a given object in relation to its larger-scale relations. A *Homo sapiens* body is a stabilized pattern of trillions of cells but a part of a complex ecology.

4.15—All of these motions are happening and aggregating simultaneously. We could generalize out the point made in 4.9 and conceptualize the movement of a human body, for instance, as also the movement of fluctuations aggregated through the substance described by the unobservable Scale 0.

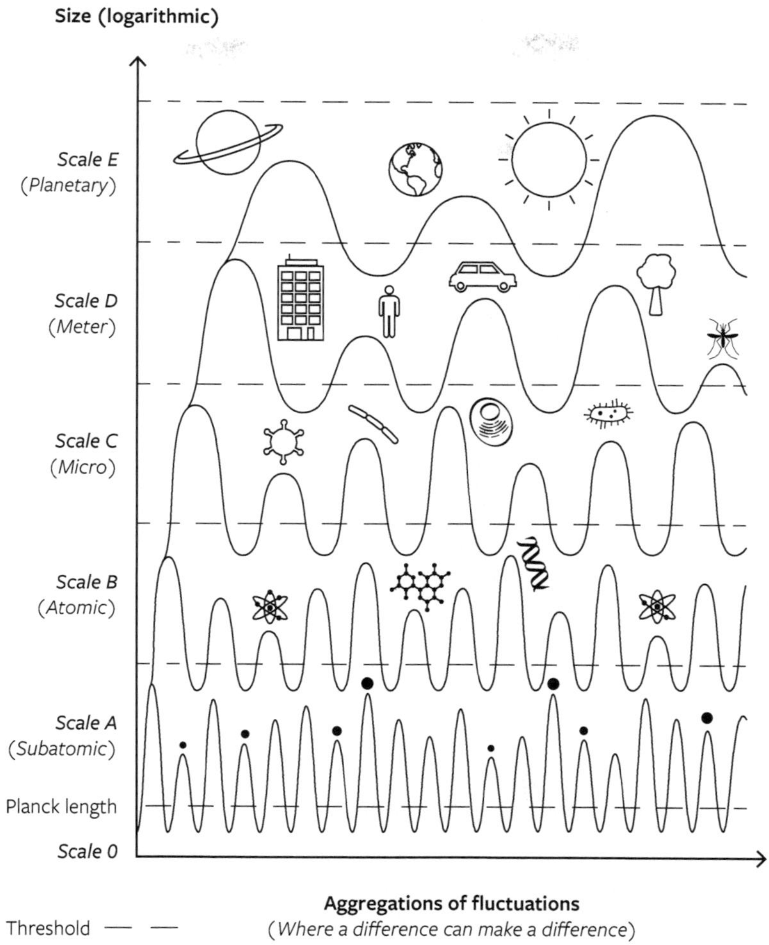

FIGURE 4: The genesis of objects: layering of objects through levels of aggregation and stabilization. If we generalize out the object form of Figure 3, then we can imagine layers in which sets of objects at any given scale are a result of aggregations of the scale below. Each parabola is an apparently discrete object at its given scale, even though it still is part of the same substance seen at lower scales. These then aggregate into objects discernible at a larger scale. Image created by Graphit Science & Art, LLC.

4.16—We could summarize the whole thought experiment as levels of aggregate effects stabilizing into objects at each level, with their own rules provided at each stabilization generated from simple compounding of effects on a lower scale (Figure 4).

4.17—Every part of substance not discernible at a given scale is, properly speaking, still there. It is just simply incapable of being registered at that scale, although it manifests in other ways as part of the patterns of the same substance. The idea that objects are separated by empty space is a product of this-scale lifeworld and vision. In turn, the idea that outer space is largely empty is a product of scales of the solar system or galaxies, where the significant objects, while themselves incredibly large, also imply large gaps between possible effects able to aggregate at those levels.

4.18—It seems that the "spaces" between objects increase in proportion to the size of those objects. This implies that aggregation tends toward proximity.

4.19—Thus, aggregation might be rearticulated as cohesion, particularly when we arrive at or beyond Scale B. Aggregation would be cohesion if considered in terms of how the things at Scale B are able to come together to have effects at Scale C. These points (4.17–19) might be essential for continued attempts to understand what the various "forces" of physics are describing, such as gravity.

4.20—This thought experiment should not be interpreted as reductionist. The experiment here is about the generation of discernible objects such that it produces objects with seemingly empty "space" between them that serve as discrete entities for observation and interaction at any given scale. It is not an argument about causality except to suggest that it is likely that the parameters (degree of freedom of action and existence) on any scale might turn out to be functionally equivalent to the principles of aggregation on the next smallest scale (see 6.21–6.34). The experiment does identify perhaps the most productive element of reductionism: that it has produced the impetus for defining the ground rules for aggregations of effects.

4.21—This thought experiment does highlight the persistent role of probability in science, an aspect central to the development of quantum physics and in the general applicability of systems theory and nonequilibrium thermodynamics.[6] Probability here is not just a function of time but is a function of the aggregation across a scalar threshold (see 6.21).

4.22—We can thus make a more general hypothesis about science's mapping of the regularity of nature: science makes use of the stability able to be defined out of scalar aggregation of effects in order to identify ways that reality is predictable and follows certain functional laws (see 6.39).

5

From the Top

The Second Thought Experiment in Scale

5.1—If scale can be conceived as a systematic alteration in the resolution of perspective (1.15, 3.44), then we can, as a thought experiment, pinpoint a spatial perspective as a function of a coordinate that is a division—or, more precisely, point of specification—of this One, which also encodes the degree of detail or resolution provided by the coordinate.

5.2—Let's designate the largest, most inclusive possible perspective of the universe as 1 (see 1.23, 3.38, 4.2). This may be the observable universe, but it need not be. If we should discover more in the universe, then we can, in the style of Hilbert's Hotel (see 3.41), move the 1 one decimal further up so that what was previously conceived of as "one" simply becomes the first decimal (division) of this new "one." This maneuver should function as long as any new discovery can be said to be "larger" and not of a different sort or alongside what is (although it remains to be seen what such metaphors, as in the notion of parallel universes, might actually mean, particularly since such theories persistently use metaphors from nonscalar experience, as is the case already in "alongside of").

5.3—If the largest conceivable scale is 1, then we can specify a location within this 1 as a fraction of 1.

5.4—In doing so, every additional decimal is not a higher degree of accuracy (as it is usually conceived in mathematics and engineering) but rather a kind of zooming in. Thus, if we conceive of 1 as a sphere, then 0.243 would specify a location within that 1 that would appear as a kind of resolution (1.15) of that 1. If we added four decimal places, 0.2438531, we are increasing the resolution while in some sense staying in the same location. One might view this fraction as a kind of sphere within the 1 whose size is specified as a fraction of that 1. As one adds to the decimal, the sphere shrinks to specify a smaller portion of that 1.

5.5—This notion clarifies the change of perspective entailed by scale. In "resolving" the decimal or, if you'd prefer, in the fractionation of the 1, one is selecting out of the One a portion of it to examine at a particular level of detail. (Note that there is no "outside" the 1 here [see 1.23, 2.27, 6.44] but rather a specification of how whatever is is being viewed within that 1.)

5.6—With this schema, we can conceptualize the way scale provides layers of existence within the same "thing." At 0.285038472 one might observe a galaxy, but then at 0.28503847234130979123236719 one might observe a cow.

5.7—This experiment clarifies how to think of scale as resolution. The decimal describes an alteration of perspective, which changes what you are able to observe at any given degree of specification.

5.8—If we apply this to what is currently observed within the scale ranges of the cosmos, we can specify the scale domains as divisions of this 1. First, using the meter as our reference point (see 2.14), we can specify the relative size range of major objects (Figure 5). We can then convert these differences into a chart for the scale domains (defined at 3.35) by simply counting down from the largest size, in this case 1×10^{26} meters, so that we have a decimal correlating with these differences in size. We could then chart the resolution 0.2516831261814905484783968731932660873193964038 as in Figure 6.[1]

5.9—This conception permits us to specify thresholds with some degree of accuracy by considering when and where a logical typing in objects actually occurs, as is already labeled in the third column in Figure 6. Here we can see how scalar thresholds are made possible by their logarithmic nature. Because we are dealing in orders of magnitude, we could think of thresholds as singularity points for objects of relevance on each scale.

5.10—As noted in 3.37, there are some transitional ranges where one might place a phenomenon in one or the other threshold depending on whether we privilege one scale or another, such as geological features.

Observable Universe (diameter)	**8.8 x 10^{26} m**
Clusters of Galaxies	**1 x 10^{24} m**
Galaxies	
Gigantic	1 x 10^{22} m
Mid-range	1 x 10^{20} m
Dwarf	1 x 10^{19} m
Solar Systems	
Oort cloud	7.5 x 10^{14} m
Heliosphere (of our sun)	2.6 x 10^{13} m
Space Objects	
Large stars	5.5 x 10^{10} m (Rigel)
Small stars	6.9 x 10^{8} m (our sun)
Large planets	7.1 x 10^{7} m (Jupiter)
Small planets	4.8 x 10^{6} m (Mercury)
Continents	1 x 10^{5} m
Objects	
Landscapes	km (1 x 10^{3} m)
Animals/plants/objects	m
Insects/small objects	mm (1 x 10^{-4} m)
Microbes	
Single cell	100 μm (1 x 10^{-5} m)
Organelles	7 μm (1 x 10^{-6} m) (nucleus)
Virus	100 nm (1 x 10^{-7} m)
Molecules	
Macromolecules	10 nm (5 x 10^{-8} m) (average protein size)
Molecules	2 nm (2 x 10^{-9} m) (size of DNA molecule)
Atoms	0.1 nm (1 x 10^{-10} m)
Subatomic	
Atomic nucleus	1 x 10^{-14} m
Protons/neutrons	1 x 10^{-15} m
Electrons	1 x 10^{-18} m
Quarks	1 x 10^{-19} m
Planck length	1 x 10^{-35} m

FIGURE 5: Major objects and their sizes. These are estimates that often fall within a limited range; what is given here should provide a general schematic. For an ongoing compendium of these comparable sizes see https://en.wikipedia.org/wiki/Orders_of_magnitude_(length).

Given this situation, it may be worthwhile to distinguish a kind of medial zone where one observes what might be described as the accumulated objects of a lower scale or aspects of a higher-scale object. These would be: components of a galaxy, patterns or sections in a solar system (e.g., an asteroid belt), parts of a star or planet (e.g., landscape features, continents, ecosystems, geological features, ocean patterns), organs or object components (e.g., rock patterns), organelles or macromolecules, and the ranges of forces or orbitals in or around an atom. However, *the major point of defining thresholds is to attend to logical typing as a function of levels of resolution, not the particular objects of relevance* (3.6, 3.30). Thus, macromolecules may be as complicated and large as a virus and therefore find themselves in the single-celled range. This is no problem: if these macromolecules form big enough chains they can become an even larger object—a rock or a diamond. It is not necessarily the kind of object we're trying to highlight but the thresholds of interactability.

5.11—This mode of specification fits with what is called "scale analysis" in various science and engineering fields. Scale analysis is a method of approximation that simplifies potentially complex equations by eliminating sufficiently small variables. When studying a process or attributes of an object that is primarily definable at one scale, one can successfully eliminate variables beyond a certain point and find that your calculation is perfectly adequate. This thought experiment suggests that a broader conception of scale analysis (perhaps more adequately called *scale-specific analysis* or *principles of specification*) is needed as a general method.

5.12—This experiment translates some problems in mathematics into the context of physical systems. Most importantly, it creates a logical typing of sets along a continuum that is tied to size. This physical logical typing carries with it the same problems already developed by Russell (see 3.6), particularly the warning that one must specify the logical type (i.e., scale) of description and be careful when creating rules or defining relations across these logical types (3.32).

5.13—Undoubtedly, this experiment has some implications for geometry, topology, and questions of measurement. While this pushes far beyond the limits of knowledge of the present author, we can point to some relevant issues by considering how one might create a kind of inverted Cartesian plane by converting the decimal into a distance from the edge of the 1. Doing so adds to the artificiality of the experiment by requiring a number of things that might not be appropriate. First, we have to posit a homogeneous space and, if we are to make it simpler, that the 1 is the same length in all directions. Then, we have to designate an edge from which to start the measurement. We must then define coordinate directions from which to start the variables, starting arbitrarily in one

<table>
<tr><th>Example Division of the 1, by Digit</th><th>1 x 10^x m</th><th>Scale Domain</th></tr>
<tr><td>0.2</td><td>26</td><td rowspan="4">Observable universe/ groups of galaxies</td></tr>
<tr><td>5</td><td>25</td></tr>
<tr><td>1</td><td>24</td></tr>
<tr><td>6</td><td>23</td></tr>
<tr><td>8</td><td>22</td><td rowspan="6">Galaxies</td></tr>
<tr><td>3</td><td>21</td></tr>
<tr><td>1</td><td>20</td></tr>
<tr><td>2</td><td>19</td></tr>
<tr><td>6</td><td>18</td></tr>
<tr><td>1</td><td>17</td></tr>
<tr><td>8</td><td>16</td><td rowspan="6">Solar systems</td></tr>
<tr><td>1</td><td>15</td></tr>
<tr><td>4</td><td>14</td></tr>
<tr><td>9</td><td>13</td></tr>
<tr><td>0</td><td>12</td></tr>
<tr><td>5</td><td>11</td></tr>
<tr><td>4</td><td>10</td><td rowspan="7">Suns/planets</td></tr>
<tr><td>8</td><td>9</td></tr>
<tr><td>4</td><td>8</td></tr>
<tr><td>7</td><td>7</td></tr>
<tr><td>8</td><td>6</td></tr>
<tr><td>3</td><td>5</td></tr>
<tr><td>9</td><td>4</td></tr>
<tr><td>6</td><td>3</td><td rowspan="7">Organisms and objects</td></tr>
<tr><td>8</td><td>2</td></tr>
<tr><td>7</td><td>1</td></tr>
<tr><td>3</td><td>0</td></tr>
<tr><td>1</td><td>–1</td></tr>
<tr><td>9</td><td>–2</td></tr>
<tr><td>3</td><td>–3</td></tr>
</table>

<table>
<tr><td>2</td><td>–4</td><td rowspan="6">Single cell/microbial</td></tr>
<tr><td>6</td><td>–5</td></tr>
<tr><td>6</td><td>–6</td></tr>
<tr><td>0</td><td>–7</td></tr>
<tr><td>8</td><td>–8</td></tr>
<tr><td>7</td><td>–9</td></tr>
<tr><td>3</td><td>–10</td><td rowspan="5">Molecules/atoms</td></tr>
<tr><td>1</td><td>–11</td></tr>
<tr><td>9</td><td>–12</td></tr>
<tr><td>3</td><td>–13</td></tr>
<tr><td>9</td><td>–14</td></tr>
<tr><td>6</td><td>–15</td><td rowspan="6">Subatomic</td></tr>
<tr><td>4</td><td>–16</td></tr>
<tr><td>0</td><td>–17</td></tr>
<tr><td>3</td><td>–18</td></tr>
<tr><td>8</td><td>–19</td></tr>
<tr><td>...</td><td>...</td></tr>
<tr><td>5</td><td>–35</td><td>Planck length</td></tr>
</table>

FIGURE 6: Scale domains charted according to powers of 10 and correlated with a decimal divided out of a 1 at the largest scale. The first column is the example number, split up by digit vertically. These digits are correlated to a power of 10 and divided into associated scale domains.

direction and providing another at a right angle from that, and third, from a right angle in a third direction from their bisection to produce a (x, y, z) coordinate plane. One could then consider the decimal as a distance by taking the highest unit (the 1)—10^{27} meters or 1,000 yottameters (Y) if we're using the observable universe—and providing three such coordinates. If we are to keep our conception of scale in view, then we need to distinguish between the abstract point of intersection, which produces a theoretically finite point, and what is still able to be observed at that point. We can do this by conceptualizing that intersection as the center point of our sphere of observation within the 1, while the amount we resolve the decimal determines how big the sphere of observation is (see Figure 7).

5.14—This scalar space has some counterintuitive properties, which are useful for helping us continue to think in terms of scale. For one, we could not provide a different number of decimal units in the three coordinates and still adequately speak of it as providing a position, since it would resolve the 1 at different scales. The distance coordinate in each variable is not so much a point on a coordinate plane as an indication of how many of a given unit of measurement we are taking. Thus, if we're using 10^{27} as our base size, then in the coordinate 0.382057914, the first decimal (3) indicates how many 10^{26} meters, the second (8) indicates how many 10^{25} meters, and so on. One could thus convert the decimal into a number with the final decimal determining the unit of measurement: 382,057,914 exameters (10^{18} meters). This is why we have kept it as a fraction: it keeps in view that this is a kind of specification via fractionation of a whole possible length and accounts for how increased specification produces a change in what is resolved. It would be strange, for instance, to designate the distance as 382,057.914 zettameters (10^{21} meters). We can further note that the lack of additional decimals does not imply that they are 0, that is, that the number used here is 0.382057914000000000, but rather that these numbers have not been specified at all because to specify them would change the resolution.

5.15—This thought experiment provides a means of generalizing Benoit Mandelbrot's "coastline paradox," which notes that the contours of a particular object (e.g., the coast of the British Isles) does not have a well-defined length. Rather, the length depends on the scale at which one is measuring, since a smaller scale will include more variations within its

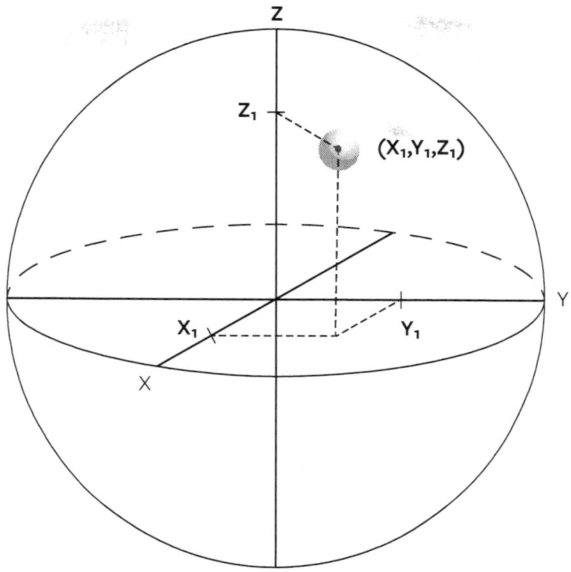

FIGURE 7: A scalar plane. The outer sphere is the theoretically posited highest point, which is set at 1 and artificially rendered equidistant in three dimensions. An (x, y, z) coordinate plane is set up using an arbitrary starting point at the edge of this 1. Unlike a Cartesian plane, the coordinates (x_1, y_1, z_1) are not given from the intersection of (x, y, z), which is an abstract point, but from the edge of the 1. These might be described either as distances from the edge of the 1 or, more to the point, a decimal as a fraction of the potential length of it. In turn, (x_1, y_1, z_1) is not an abstract point at their intersection but a sphere of observation as large as the decimal is resolved. Image created by Graphit Science & Art, LLC.

measure.[2] When put in terms of this scalar coordinate plane, we find a generalized coastline paradox for the measurements of objects and processes. Thus, if I wanted to find the boundary of an object, say a planet, and I resolve the decimal to the meter length, I may define this boundary differently than if I resolve it to the terameter (10^{12} meters). In fact, it is unclear the extent to which we might say that this same "point" we call the "boundary of a planet" is even sensibly described at a terameter. The coastline paradox implies that one might find smaller units to describe the coast, but if one were to select too large of a unit then it would not work as a measure at all. Thus, if you measured the coastline of the British Isles with a megameter (10^6 meters), then the measure would exceed the length of the object. The attempt to divide that unit would actually change the unit of measure, effectively changing scales. Something similar must happen in smaller units of measure: does it make sense to measure a coastline at the scale of the nanometer? Would a clear boundary be possible at that scale?

5.16—The implication is that we need to match the scale of measurement to the scale of the process in order to adequately study its properties. In practice, we might want to take reference to an object on one scale but measure it on another, but this does not change the domain of relevance for a measure or observing apparatus. If algae are growing along a coast, then do we have a coast measured at a micrometer scale but defined at a meter scale? If we specify more carefully what we're talking about, then we would have to say that the micrometer is relevant for each individual alga, but for the aggregate algae the meter is relevant. So we have, again, the problem of relevance as a function of resolution (1.15, 3.28).

5.17—We could attempt to conceive of time in relation to these spatial coordinates. To chart time, one would first have to select a range of perception tied to a particular location, observe an object at that range of perception, and then track its progress toward a second point. In this sketch, we can posit here that timescales correlate to spatial scales partially because timescales are already built on the logical typing of spatial scales. Thus, the scale of the planet corresponds to timescales of relevance for that large of an object. (If we consider this point in relation to our first thought experiment in scale, we might say that fluctuations also equal a kind of time range out of which differences might make a difference.)

5.18—Surprisingly, this thought experiment does posit the *Homo sapiens* as somehow in the "middle" or center of the cosmos in two ways. If we use the "observable universe" as the boundary for our coordinate plane, then this places the observer at the center of possible existence within our scalar coordinate plane (and theoretically someone somewhere else in the universe would have a different starting point for that boundary). However, in a scalar sense we find that the meter is around the center of possible sizes of observation. This is not necessarily a privileged aspect of humans, but of the meter scale more generally: it may be that, just as the micrometer seems to be the appropriate scale for life to begin, the meter scale is the scale at which a certain kind of complex behavior and cognition is made possible.[3]

6

In the Scalar Simulation

The Third Thought Experiment in Scale

6.1—Imagine that reality was capable of being fully represented in a virtual simulation in which one could pause all motion and zoom in and out much like you can with Earth-modeling software, except with as much detail as reality itself seems to produce. (This is obviously impossible. Specifying the ways that this is impossible as we proceed will tell us as much as the experiment itself.)

6.2—In such an interface, you could have a five-dimensional experience. In addition to the three spatial dimensions and time, one could alter the scale of observation. Thus, one would have five controls: up/down, left/right, forward/backward, time forward/backward, and scale larger/smaller. We can think of this experience as being inside the 1 specified in 5.2. We could even track the coordinates of location via the coordinate system developed in 5.13.

6.3—As per 5.17, for the interface to be navigable, our space and time controls must be scaled as well, so that, for example, at the scale of the nanometer, one moves at the unit of nanometers and in a time range relevant to behavior at that scale.

6.4—Moving through this interface on the scale of the meter would be like our normal experience of navigating this world. But the point is that the interface would be scalable such that one might zoom in and out in a way that would alter the scale of this three-dimensional space.

6.5—Here we could experience more acutely the transformation of the entirety of the cosmos simply by altering the scale of observation. If you begin focused on a person, at the meter scale you could examine around them, you could even bisect them and see organs and parts, but you would not see cells. When we change the scale to the micrometer, a refocusing—that is, a change in resolution—has to occur, otherwise one would only see big stretches of, for example, surfaces of the skin. But what has happened when you arrive at cells? Can you see the object "body" anymore? Here we find the full generalization of the point made in 1.15: no, the body is no longer visible. Not even organs would be visible except perhaps as what might be deduced from the patterns of the cells in relation to each other (e.g., the curvature of the cells in a blood vessel), the presence of certain larger nonliving parts, or by similar types of cells.

6.6—This is not, properly speaking, a "zoom" in the sense of cinematic zoom, which is a result of magnification. Magnification alone will not produce scale, since scale requires a change of resolution.[1] In our hypothetical interface, this is the difference between moving forward and changing the scale. We can add, in response to critiques of the "smooth zoom," that the change in scale here is smooth only in the sense that the change in scale can be placed on a continuum.[2] However, to produce our "controls" that can change scales, we require not a linear change but a logarithmic one, as per our chart in 5.8 (Figure 6). In addition, even if the change in scale is smooth, the crossing of scalar thresholds is clearly not smooth. Rather, a smooth or continuous change in scale produces a discontinuity in observation.

6.7—In this experiment we will speak as if the interface could be produced in visual terms most familiar to *Homo sapiens* viewing at the meter scale. This is a significant artifice, since it is unclear how much one might render these domains visible. Of course, light is impossible to use beyond a certain scale, and this interface would have to translate other forms of differentiation into visual form to produce these interfaces. In doing so it would highlight isolation, distinction, and separability—aspects most clearly present in vision at the meter scale.

6.8—The idea of bisection mentioned in 6.5 assumes an additional artifice: that we can move through an object that would otherwise be solid. The difference between going "in" in the sense of bisection and "in" in the sense of scaling down highlights a difference in scalar thinking: to go "in" is not to go to a smaller scale.[3]

6.9—Let us now experiment with using our interface to try to understand something about reality. We can consider three different sorts of questions that scientists might ask and see how we might use our scalar interface to answer them. First, what is X? Second, what causes Y? Third, how does Z work, what does Z do, or, more generally, what will happen next? The first is a question of properties, the second of causality, and the third of predictability and process.

6.10—We must first highlight the strangeness of observing the scalar transformation of objects not as a function of time but rather as a function of a change in spatial scale. *How have we changed all of reality without taking any action within it?* This aspect is why I have avoided speaking of scales of time as prior to spatial scales. In approaching these three questions with our interface, let's attempt to avoid time at first to see how scale operates without any action other than the change of the level of observation.

6.11—For the first type of question, we can select a scalar object: What is a cell? To ask this question, we must first switch scales to the micrometer (or slightly smaller), where the phenomena of cells can be resolved. We could then pick out a cell and, without changing scales or time, observe that the cell had certain definable attributes, including cell walls and various organelles. Equipped with a grade school chart of a cell, we could go through the cell and identify its various components.

6.12—Finding a coherent definition for a cell should not be an issue as long as we remain on the same scale. After all, the cell has some clear boundaries at the level of the micrometer, and certain components that make it appear as a separable object. However, if we switch scales, then the object of study will no longer appear like a coherent object. We can experiment with this as a general rule about scale: *studying something as a clearly defined object is only possible when we remain at one scale.* Thus, while we used scale to find the object "cell," we find that we cannot easily add a scalar shift back in and keep in view this object for study.

6.13—There are a number of ways that a seemingly separate object might appear not separate within our simulation. Without time, for instance, things might be indistinguishable if there is no discernible space between them until, adding time, they move apart. Moving our simulation forward and backward in time would produce a sense of objectness by permitting things at a given scale to demonstrate their distinctiveness through their ability to move as a unit at that scale.

6.14—Once one has resolved and defined the cell, it may be possible to study the cell from the perspective of the organism or the molecule. Doing so requires that we switch back and forth between the two scales to see how they relate. The result is quite productive: we might be able to see that the cell is a heart cell by noting its location in that organ. We can then specify attributes of a heart cell as opposed to cells belonging to other organs. But doing so requires that we switch scales so that we can discern the relevance of this organ structure to the larger-scale organism. We might be able to acquire something of a distinction between heart cells and bone cells by traveling in the same scale and noting a difference in kind, but only with a switch in scale would we be able to understand this difference as a function of an organizing principle of relevance on another scale. The term "heart cell" thus already refers to the same object from the perspective of two scales. More precisely, "heart cell" categorizes an object observable at one scale (cell) in terms relevant to another scale (heart).

6.15—Could we distinguish a symbiotic microorganism from a body's cell when remaining on the same scale? This question exposes our attempts to define a system or class of relevance using a larger scale (the organism). If the microorganism is an essential part of the system, why do we call one a cell and another a microorganism (e.g., why is a microphage a cell but a bacterium in your digestive tract not?). One answer is to use DNA, but DNA requires a switch in scales; plus, it does not account for the well-known fact that mitochondria contain their own unique DNA. The point here is to highlight how certain concepts and distinctions can drift from one scale to another and thereby present problems for scientific description.

6.16—We may also study the cell as it is formed by molecules. Rather than merely a strong reductionism, this maneuver has great explanatory power for understanding properties apparent at the level of the cell. Molecular biology is able to, for instance, study the properties of proteins or DNA in order to say a good deal about what a cell is and how it works. But note how this replicates the same pattern we just defined in 6.14 and 6.15; molecular biology effectively selects new objects primarily discernible on a smaller scale (DNA, protein molecules) and studies those in relation to this larger-scale system.

6.17—These last few points (6.14–16) consider only the immediate scale domains. Some scientists study cells in relation to ecology (e.g., the effect of bacteria on the planetary ecology) or study ecological factors in relation to cells (e.g., the role of the great oxygenation event in producing eukaryotes). In these cases we have significant shifts in scale being deployed to say something about what an object is. Indeed, once an object of study gets defined on any scale, we can relate it to objects and processes on any scale—as long as we remember the scale at which each object is defined.

6.18—For these observations (6.11–17) we chose a living object. If we apply these observations to nonliving objects, other aspects arise. Consider the question "What is water?"—or, to place the question on a particular object in our simulation, "What is this water in this glass?" What scale are we to select to answer this question? At the scale of the meter, water in our simulation would include everything within the glass; we would not have the grounds to distinguish the water from its contents (e.g., one is reminded that H_2O does not actually conduct electricity). At the micrometer scale we could distinguish the water from any microbial and any large macromolecular components. Only at the nanometer scale could we distinguish the water molecule from other molecules within it. We could then study the attributes of H_2O, with the understanding that water's distinctiveness arises at the level of the molecule. We can then observe, for instance, that water at the scale of the planet exists as clouds, oceans, and glaciers and draw some relation among them.

6.19—We can now continue to develop the concept of thresholding (1.9–10, 3.27–29, 4.8, 5.8–9) and speak about phases of materials. Clouds, oceans, and glaciers are three configurations of the same molecular structure (H_2O) relating to each other in a configuration that makes them visible and effective at the meter and kilometer scales. In our interface we can see the significance here of phases, since the difference between ice, water, and water vapor might be described by switching scales and observing the relation between H_2O molecules at the nanometer scale that changes how they appear and what they are able to do at the meter scale. As systems scientist Ricard Solé notes, there seems to be a scalar nature of phase transitions along a critical point: "the term *critical point* . . . describe[s] the presence of a very narrow transition domain separating two well-defined phases, which are characterized by distinct macroscopic properties that are ultimately linked to changes in the nature of microscopic interactions among the basic units."[4]

6.20—Combining 6.18 and 6.15, we can add that the distinction between living and nonliving objects depends greatly on the scale of observation. Because life is invariably a system of smaller-scale objects coordinated into an object on a larger scale, what is considered living depends on the scale used to define the living system. Thus, at the nanometer scale, nothing is living unless we want to consider molecules involved in a cell or microbe as themselves living (then we run into the problems of definition found in 6.15). The point becomes more important at the micrometer scale: in an organism we distinguish between cells and microorganisms as living and material structures (e.g., calcium deposits that make bones) as nonliving. But if we were to change scale to the organism (meter), these "nonliving" components would be an essential part of an organism. Following 6.14, we might say that these nonliving structures seen at the micrometer scale are living in reference to the scale of the meter. This point becomes more relevant in the life/nonlife distinction between the meter-planetary transition, since many things we habitually treat as not part of life—or as mere materials *for* life—ought to be properly treated as part of life on the scale of the planet. This point reiterates the significance of the notion of the biosphere and Lovelock's notion of Gaia as attempts to conceptualize life on the scale of the planet (3.29, 3.32).

6.21—We can now turn to our second form of question, "What causes Y?" While this question seems invariably about the result of action in time, we want to highlight how easily answers to this question can be posed in terms of scale. If we freeze time and stay on the same scale, what can we say about causality?

6.22—The great use of our interface is that we can use it to find a specific answer to a defined problem for the sake of intervention. Medical examples are the most relevant, since they include a presumed problem to locate and fix. Let's say that I have a terrible headache and a pain in my back. Let's freeze time and consider what we can do to examine this situation in our interface. We can check for particular causes without changing scales: there might be someone sitting on my back (observing the relation between two objects on the same scale), or I might be able to peer "into" the body without changing scales and identify whether or not there are any fractures or swelling in the spine (observing attributes in objects on that same scale). If we find no fracture (or someone sitting on my back) but just swelling, we have not found a cause of the malady but a symptom of it. At this point it would not make sense to scale up, since that would leave behind the object (the body) that is manifesting the symptoms. Instead, we scale down and examine the cells in the spine, only to find that there are *Neisseria meningitidis* bacteria present in the spine. Thus, we have found a diagnosis, cause, and potential intervention by going to a different scale.

6.23—Arguably, Pasteur's great discovery was to pinpoint disease's primary cause as activity on the microbial scale.[5] With this notion in place we could always look for a microbial cause for other ailments; we could, for instance, find someone with schizophrenia within our interface and examine if they also have some kind of bacterial cause for this mental illness, as has been periodically suggested. The fringe nature of this theory points to how not all disease is relevant in this way. For instance, another condition might be caused by too much iron (nanometer scale), by other configurations of the system at the micrometer scale (hormones), by DNA sequences, by societal problems, or by patterns in neuronal configurations.

6.24—In these cases we are trying to locate differences (the bacteria) that make a particular kind of difference (the illness), which has been selected in advance. Once we have identified the bacteria as the cause, we can intervene to control that particular object of relevance on other scales: from the epidemic perspective (above the meter scale), we can instantiate quarantines, sanitation practices, and immunizations to effectively target that scalar object. In fact, the effectiveness of this intervention relies on—or even is the evidence for—the proper selection of this difference.

6.25—Significantly, "cause" here is the specification of the difference that makes a difference. If this difference (the meningococcus bacteria) is removed, the disease is likewise removed. To the contrary, if a symptom is removed (taking medicine to control the swelling), then the disease is not removed.

6.26—In actuality, this specification is quite difficult, particularly since we do not have the convenience of our imaginary scalar interface. Even if we did, something like cancer is not nearly as straightforward. How would one find the determinant factor for intervention? Even with a clear specification—cancer is an uncontrolled replication of cells—the larger question remains: what causes certain cells to replicate uncontrollably? Even if damage to DNA produces cancer, what causes the damage? Intervention here might already be of different sorts, which reflects the scale at which one is trying to locate the significant difference: one could fix the damaged DNA, remove the cells via chemotherapy, or avoid the cause of the damage in advance by altering environmental factors (working in a nuclear plant; staying out in the sun too long; smoking tobacco). In any case, the primary scale of interest may be the molecular DNA, but the chain of causality leaves open different scales of intervention.

6.27—The point here is to highlight how scale permits us to identify multiple points of causality by examining the whole scalar configuration. But one can see how there is a general preference for scaling down as a description of causality. Indeed, this could be a definition of reductionism. The caveat for moderating this tendency: *scaling smaller describes the same thing viewed in a different way as the cause of that same thing.*

6.28—In one sense, there isn't a cause at all when we switch scales except insofar as we want to identify a particular intervention or effect. Consider the psychiatrist examining mental health: does the presence or absence of serotonin cause depression? If one is depressed, these neurotransmitters are likely less present. Are they diminished because you are depressed, or are you depressed because they are diminished? This is the same brain described in two ways, via two scales: at the scale of experience and at the scale of neuronal components. However, this does not provide a necessary path for intervention. If we alter the balance of these neurotransmitters, one might effectively alter the feeling of depression while still failing to identify the organismic, experiential, social, or environmental configuration that might instantiate the depression.

Alternatively, one might fruitlessly alter social or experiential factors and find that the feeling of depression does not disappear. The person being treated for depression is always caught in the conundrum of this scalar shift.

6.29—At the same time, these questions about causality cannot avoid the question of how objects at one scale affect those at another scale. We can define two different senses of *scale effects* here. First, there is the effect that occurs when you change the scale of observation. The ecologists Jianguo Wu and Harbin Li, for instance, define scale effects as "the changes in the result of a study due to a change in the scale at which the study is conducted."[6] Second, we have the ways in which actions or objects on one scale affect objects or actions on another scale. To separate these two notions, we can keep the name *scale effects* (i.e., impact of scale) for the effect of scale on our observation, and *scalar relations* (i.e., impact across scales) to describe how action on one scale might influence action on another scale (see 3.31).

6.30—We can divide scalar relations into aggregation and conditioning. Aggregation is the way that the attributes and actions on a smaller scale accumulate into larger-scale actions and attributes (4.8). Thus, the way that molecules of carbon become a diamond or cells become a body are forms of aggregation.

6.31—Conditioning is the way that larger-scale objects or processes set conditions that constrain, enable, shape, or otherwise influence lower-scale activity.

6.32—Both aggregation and conditioning are a form of "setting conditions" for action at other scales. Aggregation sets out a possible range of configurations in a larger structure by providing delimitations in form and attributes. Thus, carbon sets different aggregating conditions than silicon. Both may be mediated by certain principles of aggregation, some of which might be principles of all aggregation across any scale (see 4.11).

6.33—Likewise, conditioning is a setting of conditions under which smaller-scale actions are able to be affected by changing what is available for smaller-scale interaction. For example, self-assembly techniques in nanotechnology set conditions for given materials (e.g., carbon) so that they will assemble themselves into a particular configuration (e.g., carbon nanotubes).

6.34—We have to be careful about how we conceptualize aggregation and conditioning as causal conditions. It may be possible to specify how certain conditions on one scale become effects on another scale, but it is a larger leap to suggest that aggregation or conditioning causes another scale's behavior even if these provide means of intervention or prediction.

6.35—Ideas of control, including notions of hierarchical power, are constantly falling sway to this problem. In what way do "I" control my hand, if I consider my hand as an aggregate of cells? It might be more accurate to say that action at the meter scale sets conditions under which this aggregate of cells is able to do the things that they do. To replicate this structure at the level of the planet: environmental factors, such as oxygen or carbon levels, weather, mineral sources, and geographic features become essential conditioning factors around which living organisms at the meter scale are able to live and organize. Human interventions at this same scale via large-scale infrastructure similarly set conditions for rather than controls for the organisms (human or otherwise) that live within them. From a scalar perspective, what we call "control" is a limited phenomenon that may apply only to same-scale interactions. More importantly, scale starts to highlight how "control" might be a narrative about causality (i.e., a way of mapping a flow of effects, preferentially written to emphasize human intervention) rather than a fact of phenomena. This, of course, has most relevance in trying to conceptualize control at the level of human governance, but it applies equally to our example of self-assembly of carbon nanotubes. Is the engineer controlling the molecules in setting the conditions for their self-assembly?

6.36—However, 6.35 does not preclude the definition of principles of aggregation and conditioning. To avoid the premature and often merely preferential narrative of control, we can note that such principles work separately from the objects aggregating or conditioning. That is, such principles are not caused by any one component but are manifest in the pattern of them.

6.37—We have unwittingly introduced time to make these points about scalar relations. But if we return to our interface and do not factor in time, then we could theoretically see how attributes on one scale relate to other scales. We could examine the structure of the carbon in a diamond to see how it relates to the sharp edges and shiny surface at a larger scale. Similarly, we could see how the heart cells are conditioned by the organ, specifically the role it plays in the larger organism. In turn, we can deduce how the larger organism sets conditions for these cells by making available certain materials and limits for growth and form, limits that one could see even in the structure of the organism itself.[7]

6.38—Although we have already implied a great deal about process and predictability, we can now move fully to this third set of questions. We should note that there are some scalar reasons—having to do with the principles of aggregation and conditioning—why we cannot answer the question "What will happen next?" without going forward in time. To illustrate, we can follow Gregory Bateson's discussion of some related scientific concepts in *Mind and Nature*. His example: if one were to throw a small stone at a pane of glass, one could not adequately predict the resulting fracture pattern. Counterintuitively, the more controlled and precise the conditions—using homogeneous, polished glass with a stone at a controlled speed—the less predictable the fracture pattern becomes. However, if there is a preexisting mark in the glass, you can predict the pattern to some degree. Out of this example, Bateson derives two principles that are principles of scale:

6.39—Bateson states that convergent sequences are predictable but divergent sequences are not. Divergent sequences "concern individuals"—including "the crack in the glass"—while convergent sequences involve aggregate behavior. Bateson's point applies to scale more generally: "The movement of planets in the solar system, the trend of a chemical reaction in an ionic mixture of salts, the impact of billiard balls, which involves millions of molecules—are all predictable because our descriptions of the events has as its subject matter the behavior of immense crowds or classes of individuals." Or, as he says elsewhere, "the generic we can know, but the specific eludes us."[8] In other words, aggregation provides predictability by bringing together the diversity of behavior from a scale below into aspects that are more regular on a scale above due to principles of aggregation (see 4.12, 4.22).

6.40—In terms of scale, we can say that breaking points and fracture patterns describe behavior at a lower scale at the point where they make a difference at a higher scale. Without time, one will be hard pressed to predict the moment of transition or pattern, since one cannot easily find the specific lower-scale item that will serve to instantiate an aggregate effect. We would have a hard time scouring our interface at the scale below, hunting for the molecules in the glass that will serve as the path for the cracks in the glass. The exception is when there is something on the higher scale that conditions this lower-scale behavior, such as the scratch on the glass. But even so, this conditioning does not give us complete information in advance about the precise flow of lower-scale events, even if it does delimit to some extent their aggregated behavior.

6.41—Once we add time, we can examine the patterns of conditioning and aggregation by observing what does happen within any object. We might see, for instance, that a breaking point is reached, know that it has to do with molecular-scale interactions, and retroactively identify the location of that instantiation at the scale of the molecules. But the same problem is always pushed forward in time: we can know something will happen but will be hard pressed to find where the threshold will occur without moving again forward in time. We will always be limited in our ability to trace out the exact flow of a phase transition, aggregation, or result of conditioning except in aggregate, after the fact.

6.42—Here we have perhaps come to the limit of the usefulness of the thought experiment without defining the limits of the whole premise. On a basic level, such an interface would have to simply *be* the universe, since any attempt to encode this complexity would have to be as elaborate as existence itself. This is most relevant for our discussion of time and predictability, since only in the hypothetical interface could one locate the molecule that forms a breaking point. In reality, the breakage itself is this discovery.

6.43—This is also a point about the great variety of existence implied by scale and about the limits of modeling. Scientific modeling will, of course, take advantage of scale to compress the situation via the aspects discussed here (6.38–41), but its limits might be more precisely defined by considering the scalar maneuvers that science uses to produce anything that can be considered predictability, intervention, and explanation. In chapter 10 we will discuss the function and limits of science as a process of specification.

6.44—This experiment's interface follows the common ruse of objectivity, imagining a viewpoint outside the interface to observe reality. But all experience is already within the "interface," not separate from it. In many respects, experience is a result of the operations such an interface would model. But this limit does not mean that this scalar map is an inappropriate way to interpret the knowledge that we do manage to stitch together in the process of science. In fact, *Homo sapiens,* whether the one called "scientists" or otherwise, is capable of prodding at and mapping these scalar layers of reality precisely because they are a part of it and within it.

6.45—However, in eliding the embeddedness of observation, our interface does not account for perspectives had within or by parts of reality. How does a cell differentiate the scale of the micrometer such that it can respond to events at this scale (the basic maneuver of cybernetics)? This point is essential, since we have already assumed in many of our examples that a coherent world of interaction is able to exist at any given scalar level. Because scale is about the mode of observation as well as the possibilities of interaction, one could not avoid considering how these perspectives function (see chapter 8).

6.46—When we spoke of modes of generating visualized perceptions in the interface (6.7), we were trying to make the interface comprehensible by imagining it in visual forms, which is clearly impossible. Such a visual form tied to the *Homo sapiens*'s body would also not account for how the same thing might look differently depending on what signals are received by it. If we add this observation to 6.44–45, we arrive more properly at the notion of *Umwelt,* or lifeworld, as a term describing how any organism selects out parts of reality in a particular way so that it creates a particular environment to respond to.[9]

6.47—To understand "what comes next" one would need to understand both cybernetic responses and the contours of an *Umwelt,* since the responses in the scalar interface require a responsiveness to configurations that are not always present as "objects." One might experiment with stating this in a scalar fashion: if I were to locate the way that a nervous system will respond to a stimulus, I would still have to relate any given neuron to the whole configuration of other neurons in order to get a pattern that we generally call a "learned response" or, in computer terms, a "program." Indeed, in computers we see how a program is to some degree a scalar system: it uses the states of its components to build a pattern capable of determining a response. One would be hard-pressed, even with a great amount of effort, to discover in our interface how such a system would respond next without adding in time—that is, by just running the program.

6.48—The notions of pattern and pattern recognition are clearly far more important in a scalar conception, since scale produces patterns out of which one might create rules for both delineation of objects and predictions about their interaction. In this case, pattern is not simply what comes next but the relationship between the larger and the smaller.

6.49—As a final note: within such an interface (or through the systematic examination of scale via science and careful theoretical delineation) we can add three kinds of scalar principles to scale effects (6.29) and scalar relations (constraints and aggregations [6.30–33]). First, principles of how differences manifest and accrue within any scale domain. These would be *scale-specific principles* that might not be generalizable to other scales and would therefore need to be situated within the scales of relevance. As science writers Joel Primack and Nancy Abrams note, "physical laws that apply at one scale do not cease to be true at other scales: they merely cease to matter."[10]

6.50—Second, *isomorphic principles or characteristics* are patterns or similarities that can repeat between shifts of these scale domains. These provide grounds for making analogous comparisons between scales, such as learning about how to organize a city by looking at a cell.[11]

6.51—Finally, the principles of scale itself: the conditions under which we use this apparatus, thinking as carefully and broadly as possible about what this simple observational accounting does to our conception of the cosmos. Clearly, this is the aim of this work.

PART II

Configurations for a Theory of Scale

The way of thinking developed in the algorithms leaves open many questions. While the implications of scale are by no means fully clear and apparent, much resistance toward these conclusions may arise from the strangeness of thinking and speaking in terms of scale as it pushes us beyond the realm and structure on which human experience is built. How do we reconfigure our sense of reality if we are to accept these different scales of existence and the apparatus of scale itself?

We need to dwell on the configuration of reality provided by scale as it reworks three basic philosophical notions: objects, subjects, and relations. The bases for the conclusions presented in this section are already stated in some form in Part I. But because scale implies such a significant shift in our way of understanding reality, we have to move carefully through these new configurations.

While much of what follows may seem abstract, the approach here is meant to be experiential. If one adopts the scalar perspective outlined in Part I, what is the experience like? How does one reconcile this perspective with the concepts, assumptions, and ways of speaking built from a nonscalar perspective? How does one resolve the tensions inevitably faced in this reconfiguration? To this end, each chapter will begin with a manifestation of scalar experience in cultural narratives to remind ourselves that these are reconfigurations we all face in attempting to adjust ourselves to our scalar situation.

These three chapters are where I am more properly "scaling theory," in the sense of running theoretical articulations through scale in an attempt to reorient our way of talking about objects, subjects, and relations. The exegesis here often focuses on theoretical texts, drawn from various theoretical humanities and sciences of the last few decades. In addition, we will face more directly the drift between mystical and scientific ways of thinking and speaking, which we introduced in the Introduction.

These reconfigurations occur simultaneously, so there will be portions of each chapter that could be placed in the others. Although the chapters build on each other, the reader is welcome to read them in any order.

7

In-formations of the Whole

Scalar Configurations of Objects

> We are being processed along, and as we go we are changed and informed; there is no ontology for us, no concrete being.
>
> What lies hiding within each object? A garden, so to speak: the enchanted garden.
>
> —Philip K. Dick, *The Exegesis*

CUE THE SCALAR INVASION

In February 1974 the science fiction author Philip K. Dick experienced what he later described as a massive transfer of information from an alien entity. A combination of sodium thiopental (for oral surgery), a fish sign, and a flash of pink light began a series of visions that lasted for roughly two months.[1] Dick would spend the rest of his life attempting to understand this "2-3-74" experience, writing thousands of pages of philosophical and theological speculation that he called his *Exegesis.* This text, which is only recently being transcribed and published, is full of endlessly fractal theories integrating a diverse history of thought as Dick voraciously sought to understand 2-3-74.

At the core of the *Exegesis* is the Vast Active Living Intelligence System (VALIS), one of his many names (alongside Logos, Zebra, Noös [Mind], and God) for the entity beaming him information. Dick puts much emphasis on this "Vast," this sense of something larger than himself making contact. Equally interesting is his invocation of Valis as "living information." This combination—the informatic and the vast—makes Dick's experience interesting to examine in an account on scale. Consider, for instance, the following passage positioning Valis as a kind of brain:

> This vast brain must be an organizing principle. A system of linking. . . . I was taken into a thinking system . . . how, if at all, does this system exist

> independently from the constituents which it links together? . . . This model (brain-mind) is a good one for my understanding of 3-74. . . . *We* are the [physical] brain [components]. The plasmatic entity I saw which I called Zebra must have been the analog for the electrical discharges constantly moving through neural fibers. . . . So my brain, made up of millions of cells, in billions of [electrical] combinations, became one station [cell] in [of] a larger brain, linked to other "cells" [persons] . . . with Christ as the total mind [psyche]. (38:26; 354. Brackets in original)

This passage gives a taste of Dick's scalar articulation as well as its strangeness. Underlying this brain-neuron comparison is a search for a scalar view. But is Dick suggesting that the cosmos is a large, intelligent entity already, like a brain? This turn to "Christ as the total mind" is undoubtedly disarming regardless of one's religious inclinations. Why would Dick turn to such language? Or, given that his speculations also cover basic science fiction alien premises, mystery cults, and at least a dozen religious traditions, why have we run into a discursive form that seems to be, as one critic puts it, "erratic, even crackpot."[2]

Dick understood that these writings were unusual, and he incessantly doubted his own articulations. It is precisely this disorienting melding of various discourses—searching for any kind of language for understanding his experience—that makes him so useful here.[3] Dick's experience can be classed as an encounter with the strangeness of a scalar configuration of reality. His writing is a manic attempt to understand a world saturated by descriptions and experiences literally out of one's scale. Dick's occupation as a science fiction writer, combined with his predilection for asking hard questions about reality, led him to take seriously the disorientation of scalar science and the reality of his scalar experience. But he also struggled to describe this scalar configuration. Thus, just prior to the brain passage, he relates this notion of "brain" to his novel *Ubik,* which had recently earned critical praise from the literary critics Fredric Jameson and Darko Suvin (after meeting Jameson and Suvin, Dick referred to them as "the Marxists"):

> "Ubik" depicted a spontaneous generation of guiding messages from an unknown source, which was ubiquitous. The Marxists took great interest in "Ubik" and wanted to know what Ubik was. My 3-74 experience resembled "Ubik"—why? The presence of the vast (ubiquitous) brain (Zebra or VALIS) of course. In other words, in 3-74 I encountered the

> mind which in "Ubik" I called Ubik. It is the Logos, St. Sophia, but the Marxists, although evidently aware of its existence (how come?) could not accept that explanation. They may be right; a more modern, more precise formulation may be possible. I am not able to do that; I keep falling back on traditional stereotyped theological terms and concepts. (38:25–26)

In many ways, this is our situation with scale: how do we handle this reconfiguration of reality? What is the appropriate way to describe it? Why does scale find us launching into articulations that, like Dick's, seem too theological or downright crazy? What is the relationship between these theological articulations and the scientific?

This tension over "Ubik" highlights three aspects of scale that are difficult for us to process. First and foremost, we find the necessity of a highest order of abstraction—a Ubik—which leads Dick to return to theological language. What is this notion of wholeness, substance, unity, or divinity, and why is it necessary for a theory of scale? Second, Dick finds himself dealing persistently with the tentativeness of objects and division. How does this unity relate to the contingency of objects and divisions? What is the status of ontology in a scalar view? Third, he integrates this tentativeness of objects together with the language of information, tying together a condition of perception with an appearance of reality, creating something like an informational ontology. Indeed, he finds himself with one of the strangest points we came across in Part I (1.19): what does it mean to describe objects as information?

As a "fictionalizing philosopher" (*Exegesis*, 75:D-9; 693) rather than academic, Dick was unable to see how these aspects bear a fraught relationship to philosophies that were already becoming popular at the time he was writing. In both postmodern philosophy and certain analytic traditions—in what is known as the "linguistic turn"—there was a recurring push against the necessity of this highest order of abstraction, including any idea of reality as a whole, a one substance, or a transcendental signified. As Cary Wolfe notes, following Lyotard, this "incredulity towards meta-narratives" is perhaps the most characteristic aspect of postmodernism.[4] While science and these philosophies have often clashed, this attempt to, as Brian Rotman puts it, "take the God out of" various philosophies and sciences (mathematics, in Rotman's case) mirrors the general suspicion of science toward theological-sounding explanations.[5] This suspicion makes our arrival at this highest order of abstraction (first at 1.21–28) counterposed to these tendencies, much like Jameson and Suvin's incredulity toward Dick's explanation of Ubik. Yet the tentativeness of objects and the role of information are both essential to the linguistic turn.[6] Dick's statement that "a superior analogy would be to regard the universe as consisting of language" (4:166; 81) echoes a maneuver familiar in the linguistic turn. The problem for postmodernists reading Dick, then, is grappling with how this informational ontology leads him toward the very thing they find most unacceptable: Ubik. How are we to justify and explain this scalar configuration whereby objects become contingent information provided by and within a unity not divided a priori?

To explore this configuration, I will combine portions of Dick's *Exegesis* with discussions of quantum physics, atomism, ontology, cybernetics, and information theory. Following Dick's lead, we will first examine the relationship between notions of the Whole and object contingency, particularly as they arise from Heraclitus, Parmenides, and Zeno. Considering Zeno's paradox as a kind of scaling down will run us into modern attempts to understand quantum physics, particularly David Bohm's quantum theory. Responses to such holistic theories will then highlight the resistances that arise in working with this scalar form of wholeness and objects. After addressing these objections, we will return to Plato and consider how and why we find ourselves defining objects and thereby separating the many out of the One. This will reintroduce and situate a notion of "difference" within this discussion of scale and objects. It will also lead us to the bewildering formula, provided by the American philosopher Franklin Merrell-Wolff, that "substantiality is inversely proportional to ponderability."[7] With this view in place, we can then consider what it would mean to treat objects as information by examining the cyberneticist-anthropologist Gregory Bateson's notion of information as a "difference that makes a difference."[8] Finally, we will consider the results of this informational ontology not just as a concept but as an experience.

In letting this often bizarre fictionalizing philosopher take the lead in this chapter, I want to emphasize the way Dick's work rewrites these points in more personal terms. First, how would reality appear if it showed itself in this scalar configuration in which "what is" is simultaneously one and many? Second, what would happen with one's sense of reality if these divisions are now held to be tentative? Third, how does this change our relationship to these prized objects in our lives? These three aspects provide grounds for both incredulity and wacky attempts to rearticulate our relationship to reality, yet our incredulity ought to hinge on the difficulty of grappling with the scalar experience, not a balking at the strangeness of an articulation.[9] So let us see if we can find a "more modern, more precise" articulation.

OBJECT CONTINGENCY AND THE POSSIBILITY OF WHOLENESS

Scale reconfigures reality into multiple layers of simultaneously existing objects. In this way, scale literally re-dissects reality. In scaling to the small, we seem to de-compose, allowing these bodies to give way to cells, and these cells to atoms. But larger scales also entail a breakdown, not of things that appear to be separate but of the division between those things. We live among division; it produces our world of useful objects. And yet, when we consider any significant shift in scale, we see the surprising implication found in 1.21: *scale demands that this differentiation of the world has to be tentative, since each new scale revises the same things into completely different objects.*[10] An object can never be taken to exist "in itself"—even if "what is" provides the grounds for these divisions at multiple levels—since the objects themselves will change depending on the scale at which one encounters them. We then encountered the implication (1.23) that reality might be described as a Whole, since any definition of objects requires a selection of the attributes and scale at which one will define objects.

In this argument, scale has provided us with a coherent reconciliation of some of the paradoxes that are at the heart of philosophical inquiry. As the story goes, in Western philosophy Parmenides argued for the unity of being as One while Heraclitus argued for a world of multiplicity and becoming. And yet, we find in the fragments of Heraclitus this clear statement: "Listening not to me but to the Logos, it is wise to acknowledge that all things are one."[11] In this reading, Heraclitus's "everything flows" becomes a condition of the Logos as One: if everything is always changing, then what grounds does one have for saying that that there are separate things? Similarly, a dichotomy is often drawn between Vedanta's *ātman* (Self, as equivalent to *brahman,* the universal) and Buddhism's *anātman* (no-self), but we will see how a kind of emptiness of division or self might also be described as an encounter with absolute reality.

Here is the genius of Dick's 1969 novel, *Ubik*. In *Ubik* we find a perpetual decay of reality as the breaking down of objects. The premise is simple: a bomb explodes among the main characters, but then the story continues as if they all survived. However, as we proceed we learn that they might be nearly dead and cryogenically frozen in "half-life." The characters learn this through two signs: the decay of objects and the presence of messages from their boss, Runciter, directing them to a product called Ubik, which will stop the breaking down of objects. The epigraph to the final chapter includes an inscription written in transcendent terms: "I am Ubik. Before the universe was, I am . . . I am the word and my name is never spoken, the name that no one knows. I am called Ubik, but that is not my name. I am. I shall always be."[12] In this remixing of the biblical "I am that I am," Dick invokes Ubik not only as the ubiquitous but as the one substance, suggesting that it is this substance that sustains the possibility of existence (just apply Ubik, and reality stabilizes!). But the very condition for discovering Ubik is that objects are slipping away. What if comprehending and sustaining this world—even maintaining life—requires this tentativeness of objects and the substance behind it? This is what we need to unfold in terms of scale.

ZENO AND THE CONTINUUM GESTALT

We do not need scale to provide us with the sense that the world is impermanent, but it adds a new dimension to this contingency. It helps us see why Heraclitus's maxim fits with Parmenides's argument that "what-is is ungenerated and imperishable; Whole, single-limbed, steadfast, and complete."[13] Dick cites Parmenides and Heraclitus throughout *The Exegesis* to highlight this contingency of existence. He suggests that Parmenides was "the first philosopher to prove beyond doubt that what our senses perceive as the Real World *cannot* in actuality be real" (4:163; 74) while also pointing to Heraclitus's suggestion that people live unaware, as if they are in a dream (e.g., 5:43; 140).[14] Dick's musings on the Sanskrit term *maya* make this same argument; this notion of the grand illusion is much misunderstood and maligned, but what if it is simply a description of the contingency of objects? Scale launches us into the same possibility from another direction.

Dick provides us this direction through his playing with Zeno's paradoxes of motion. His musings on Zeno can be traced back to Dick's 1953 short story "The Indefatigable Frog," in which two professors, Grote and Hardy, build a test for Zeno's paradox: "Take [Zeno's] paradox of the frog and the well. As Zeno showed, the frog will never reach the top of the well. Each jump is half the previous jump; a small but very real margin always remains for him to travel."[15] Here is how the experiment unfolds: Grote and Hardy place a frog in a chamber that shrinks anything in half every time it travels halfway along the chamber. The frog hops along, getting smaller and smaller, until it disappears. Suspecting foul play, Grote examines inside the tube himself, but then Hardy closes the door, trapping Grote within. Grote heads through, shrinking until he is so small that he has to jump from atom to atom until he falls between them and out of the tube.

Dick's retelling of Zeno's paradox puts it in the form of a physical encounter with the very small. The condition that tumbles Grote out of the tube is that he could no longer locate any object on which he could move within the chamber or be contained by it. The result: the division process itself takes him outside the apparatus that is meant to divide him. Indeed, there is something interesting about arriving at the atom, the name that was used by Democritus to designate "that which cannot be divided" but was only tied to the familiar scalar objects in the nineteenth century. The twentieth century taught us, in rather horrific fashion, that these objects were not, in fact, uncuttable. In fact, quantum physics arose from the possibility that there are distinguishable quanta within an atom that can be used to designate further entities (electrons, protons, neutrons, quarks, etc.). But the question remains: can this discovery of smaller and smaller objects continue? Do we really find more objects to leap between at these new scales?

Of course, here we are entering controversial ground. On some level I sympathize with Erwin Schrodinger, who in his final book, *My View of the World*, presents arguments for Wholeness from a Vedantic perspective,[16] states in the foreword that he doesn't discuss the conundrums of quantum theory (QT) there, because he "does not think that these things have as much connection as is currently supposed with a philosophical view of the world."[17] At the same time, QT does seem to inaugurate a discussion about the insufficiency of certain principles derived from a largely nonscalar conception of objects. Most importantly, it highlights the preference for speaking of things as objects separate and definable in themselves. When we attempt to scale smaller than the atom, objects are first presented in a primarily mathematical sense, which creates questions about their ontological reality and the epistemological conditions under which they are ascribed certain properties (e.g., whether they are a wave or a particle). *In retrospect, these observations apply to any scalar observations,* but quantum physics forced many scientists to ask these questions in a new way.

This is one way of making sense of the infamous paradoxes of quantum physics. The problems are partially about whether or not an atomic particle ought to be considered a clear and contained object. Since scale has already launched us into this same territory, there is no need to rehearse the paradoxes here; rather, we can rely on the quantum physicist David Bohm, who describes the implication as follows:

> The quantum theory shows that the attempt to describe and follow an atomic particle in precise detail has little meaning. . . . [An atom] can perhaps best be regarded as a poorly defined cloud, dependent for its particular form *on the whole environment,* including the observing instrument. Thus, one can no longer maintain the division between the observer and observed (which is implicit in the atomistic view that regards each of these as separate aggregates of atoms). Rather both observer and observed are merging and interpenetrating aspects of *one whole* reality, which is indivisible and unanalyzable.[18]

We can see that this is what Dick stumbles upon in "The Indefatigable Frog": in order to continue to divide, Grote must stay within the apparatus designed to divide. But that apparatus itself is inseparable from the process of division, and the reality it produces is "dependent for its particular form on the whole environment."

While the new divisions produced by physics at the quantum scale are highly successful and useful for many practical operations, the status of these objects as clearly distinct and localizable is, just as with any object on any scale, neither apparent nor necessary. For Bohm, the quantum scale more directly highlights this fact:

> Each relatively autonomous and stable structure (e.g. an atomic particle) is to be understood not as something independently and permanently existent but rather as a product that has been formed in the whole flowing movement and that will ultimately dissolve back into this movement. How it forms and maintains itself, then, depends on its place and function in the whole.[19]

Here we see how we get from Zeno's paradox to Parmenides's One by way of Heraclitus's flux. In terms of scale, what appears to be a clear and discernible object on one scale is only so because of the larger situation in which it is embedded. At another scale—particularly at a larger scale—the same object will and must be taken within the larger possibilities of existence that already make the object possible. Which grounds for division will we hold to be primary? How can we hold any as primary and still admit the others?

We now face a difference in base assumptions: shall we proceed by assuming a priori division, or by assuming that all "things" are in some way undivided from all other things?[20] Zeno's paradox highlights this choice. Dick later finds, in the British classicist Edward Hussey, an even clearer articulation of Zeno's paradox that brings into focus this possibility of division:

Take two objects, say these two books on my desk. They appear to be separated and obviously distinct objects. Zeno points out, to the contrary, that in order for these books to be separate objects, a third thing must exist between them to separate them. Otherwise, they must be considered two nonseparate parts of the same thing. Alright, we say, there is air between them. Then what separates the air from the books? And so on. As Hussey states it, "If something X that is is separate from something else Y that also is, then there must be a third thing Z, distinct from X and Y, to separate them."[21] But what is Z? What separates Z from X and Y?

This translates Bohm's argument into a more general scalar operation. As we divide to smaller and smaller separations, at some point it is no longer functional to base knowledge solely on localizable entities taken as divided from their surroundings. The force of the molecules of air against the molecules of the books interacts precisely because of intertangled fields of force rather than actions of clearly distinct objects. While air and books appear distinct on our normal scale, at a lower scale it is their interaction—that is, the ways in which they are clearly part of the same thing—that is important. Thus a choice must be made: do we still consider them separate entities if they are so intimately related? Is a relation a definitive separation? In this sense we divide in theory, not in reality, for the sake of highlighting or making use of parts of the world that can be considered functionally distinct according to the scale on which we are understanding and working with them.

The views from wholeness and division are two different ways of interpreting our perceptions of reality. As Dick learned from Hussey, Parmenides distinguished two forms, which are two gestalts, or ways of organizing the world: Form I is reality—that which is (Greek: *to on*); Form II is a negative reality—that which is not (Greek: *to me on*). Form II is a contingent, fragmented worldview, a product of the ability to categorize and divide. Zeno demonstrates that, in considering our divisions of reality as real, we find paradox. Thus, Form I, true substance, is more accurately described as a unity: that which is, the One, which is Whole. Form II is produced out of a rupture in which we come to believe in and act according to fragmentation without regard for this Wholeness.[22]

Dick argues, following Hussey, that the atomists took a then-radical approach to solving Zeno's paradox. They proposed including Parmenides's Form II, that which is not (*to me on*), as part of their explanation of the world. Hussey argues that the atomists do so by taking the word for "nothing" (*to mēden*) and removing the negative prefix (*me-*) to create a new word for "(no)thing" (*to den*). The being of nothingness (*to den*) is thus different from being (*to on*). In this manner, the atomists claim that "nothingness is just as real as any 'thing.'"[23] Following Hussey, Dick argues that the atomists were working against a "continuum" model of the universe. Because the atomists could not think of a world with empty space, when they spoke of *kenon* (the void), they had to make it "existent and real" in order to have it fit within this idea of continuum. "All I can presume," Dick writes, "is that the concept of continuum must have obtained prior to the atomists or they would have simply called the void what we call it: empty space containing nothing. But this is *mēden*"—that is, they would not have even bothered to name this "nothing." However, "This is . . . what they deny it: the status of *mēden*" (53:1).

Dick suggests that this continuum view was what he experienced in 2-3-74: "There was one 'thing' that is, unity and being perceived and conceived as unity. And space (*kenon*) signified something other than what it signifies—that is, what it is—in the discontinuous matter view. The key term is Gestalt. Each thing (what *we* call 'thing') was not a thing at all; it was a part, and functioned as a part (of a whole)" (53:1). This scalar relationship is a gestalt switch in which one sees unity behind division:

> Reality—including the percipient as part to whole—is experienced as one, unitary, interested structure. . . . Reality valued this way . . . is such that "the void" (*mēden*) no longer is accorded existence. . . . Hence there *is* no void, no nothing; thus all things are seen literally to be connected, and this is what I saw that I called Valis. Not a plurality of atomized discontinuous pieces of matter but, rather, one interconnected whole. (55:1)

The literal nature of the continuum is important. Division can only be considered functional, secondary (which is why it is Form II) to a reality which is continuous first. Once we empty out this differentiation—make it really nothing—Wholeness manifests: "And this turns out finally, when all falsifications are abolished, to be the Eleatic continuum which is the ancient (extinct) true Kosmos; it is still there, but we are unable to see it. We suffer then, a negative hallucination; we don't see what is not there. We fail to see what is" (55:1).

Occasionally, individuals such as Dick find themselves experiencing this gestalt switch. The result is a transformative experience that is the basis for a definition of mysticism as it emerges in the late nineteenth and early twentieth centuries. Worth mentioning here is a text that Dick encounters as early as 1975 and finds useful for understanding his experience: Richard Maurice Bucke's *Cosmic Consciousness.* This book and term show up persistently in many conversations about religion, spirituality, and science.[24] *Cosmic Consciousness* is, in many respects, the first extensive comparative study of mystical experience. As was true for Dick, Bucke's experience of this gestalt switch, which he calls cosmic consciousness, prompts him to conduct a comparative study of similar experiences and provides a reference point for considering this mode of consciousness in evolutionary terms.

Bucke's cosmic consciousness experience occurred after an evening reading Whitman. On his way home, a "flame-colored cloud" enveloped him which he quickly realized "was within himself."[25] The experience produced a realization:

> Like a flash there is presented to his consciousness a clear conception (a vision) in outline of the meaning and drift of the universe. He does not come to believe merely; but he sees and knows that the cosmos, which to the self conscious mind seems made up of dead matter, is in fact far otherwise—is in very truth a living presence.[26]

The term "cosmic consciousness" is a connection between scale, science, and the gestalt change discussed by Dick, suggesting an awareness of the highest scale at which reality is really Whole: "Especially does he obtain such a conception of THE WHOLE . . . as dwarfs all conception, imagination or speculation, springing from and belonging to ordinary self consciousness."[27] Here we run directly into the connection between scale and, following William James's and Evelyn Underhill's citations of

Bucke, what has been designated mysticism. Is this not the major provocation of Dick's invocation of divine names? What if mysticism emerges in the description of an experience of this object contingency and the manifestation of a Wholeness gestalt?

QUANTUM THEORY LOST IN THE CONTINUUM INVASION

But what is this notion of the Whole or One? I am struck by the persistence with which any suggestion of monism, unity, or wholeness is disavowed within both academic work and popular culture (cue the signature hippie characterization: "It's all one, man!"). The same reaction usually arises when attempting to take Dick's experience seriously. As British philosopher Arthur Lovejoy once wrote, "That it should afford so many people a peculiar satisfaction to say that All is One is, as William James once remarked, a rather puzzling thing."[28]

Indeed, James, in the lecture "The One and the Many," suggests that this "One" might be nothing more than a kind of number worship, partially because "it is so difficult to see definitely what absolute oneness can mean."[29] He acknowledges that "to interpret absolute monism worthily, be a mystic," but then, true to his disclaimer in *Varieties of Religious Experience* that he is not a mystic, he suggests that we "leave then out of consideration for the moment the authority which mystical insights may be conjectured eventually to possess; treat the problem of the One and the Many in a purely intellectual way; and we see clearly enough where pragmatism stands."[30] Thus, he pushes against the notion of the One in favor of pragmatic pluralism. Insofar as considerations of Oneness or Wholeness arise today, this maneuver has become standard, partially because, in one sense, James is right: without scale it is difficult to see what absolute Oneness might mean. Scale makes this Oneness clearer and helps us see how its disavowal or avoidance serves to confuse our attempts to understand some aspects of our situation. I want to make the nature and cost of this disavowal or avoidance clear by dwelling further on quantum theory. While Dick only occasionally muses on QT, it provides us one site for considering the difficulties of developing these more modern, more precise articulations.[31]

Ultimately, you do not need QT to get to Wholeness and this view of objects. This is quantum theorist Jean Bricmont's primary critique of the frequently invoked connection between QT and mysticism: nonlocality proves Wholeness just as much as Newtonian gravity does.[32] Agreed. However, if we go a bit further into QT, particularly Bohm's theory, we can locate the ways our assumptions about objects manifest. In Bohm's case, it is essential that his interpretation is an ontological interpretation of QT. What does this mean? As Bohm explains in his final book, with Basil Hiley, the problem is that subatomic quanta are statistical and therefore incapable of providing a straightforward ontological interpretation; it is unclear what these mathematical descriptions are actually about.[33] What does a Schrodinger wave function (which is used to describe a quantum system) have to do with these subatomic objects (such as electrons)? Bohr's and Heisenberg's much discussed limitations (uncertainty and complementarity) are epistemological statements: they are about whether we can know about both the position and momentum of an electron. They provide "a limitation on the possible accuracy and relevance of our knowledge of the observed system" (14). The difficulty arises when we start speaking of the limit of knowledge ontologically, as if our (in)ability to know an aspect means that these objects do or do not have them. The problem is that complementarity is "taken not as a purely epistemological limitation on our knowledge, but also as an ontological limitation on the possibility of defining the state of being of the observed system itself" (14). This is the problem with the common way of speaking of Schrodinger's cat: the assumption that perception must occur for the cat to be dead or alive is to mix up the epistemological (something comes to know the position of the quantum entity) and the ontological (the status of the quantum particle). Bohm and Hiley argue that this is a "taking the concept of quantum state too literally" (19). Why would something (usually a human) have to observe the quantum state to give it one? This generates the most problematic (and noticeably New Age) interpretations of QT: the idea that observing makes it so. This is to take the epistemological as ontological.

An ontological interpretation of QT, argue Bohm and Hiley, would separate out these epistemological limits from the ontological in order to define what can be said about reality at this scale. However, this turn to the ontological comes with an essential caveat: it is not something *knowing* that produces the attributes at this scale but rather *the system as a whole.* At first Bohm and Hiley are diplomatic about this point, calling it the "wholeness of the entire system going beyond anything that can be specified solely in terms of the actual spatial relationships of all the particles" (58). The more radical conclusion is that this "entire system" is, in principle, the whole universe (see 352–53). In other words, the condition of moving to an ontological interpretation is that the object is not treated in itself at all, but rather as a point of consideration within the entire configuration of the Cosmos. An electron can be said to have determinable attributes only in reference to the "unbroken whole."[34]

In practice, one does not need to consider the whole universe, for reasons that "scale analysis" as a method makes clear (see 5.11): within any given scale domain there are thresholds of relevance that permit us to functionally delimit what is relevant for an object. Even though an electron in the Andromeda galaxy affects one in my hand, this connection is usually negligible.[35] Although this produces grounds for delimiting relevance, this does not mean that these objects exist in themselves. To the contrary, Bohm's interpretation suggests that such constraints are themselves nonlocal: they depend on things that exceed the object to which the attributes are ascribed. From a fully scalar perspective, this is like noting that the electrons here have some relationship with the Milky Way. Of course they do; they are two resolutions looking at the same thing (see 1.20). But attributes able to be meaningfully assigned to the Milky Way will differ from those that can be assigned to electrons.

Undoubtedly, there is more to the story than I could provide here and, as Bohm and Hiley insist, there will continue to be more "depth and subtlety of laws and processes" as we continue to understand the scales below the atomic (see 319–21). My purpose is not to recount the debate about QT but to clarify the role of interpretative preferences in handling a scalar situation. If we read Bohm's interpretation of QT as being partially about a willingness to understand the basis of an object in reference to the larger whole, then such a maneuver requires a setting aside of this basic assumption derived from nonscalar experience: that the world must be made of discrete units. For Bohm this is a habit of interpretation, which treats everything as fragmented in advance, "a largely unconscious habit of confusion around the question of what is different and what is not. So, in the very act in which we try to discover what to do about fragmentation, we will go on with this habit, and thus will tend to introduce further forms of fragmentation."[36] Bohm notes, with irony, that "in the study of life and mind, which are just the fields in which formative cause acting in undivided and unbroken flowing movement is most evident to experience and observation, there is now the strongest belief in the fragmentary, atomistic approach to reality."[37]

We often see remnants of this belief in the humanities when QT is invoked for philosophical ends. We can examine this habitual fragmentation in two recent invocations of QT: Karen Barad's agential realism and Timothy Morton's use of Bohm to describe what he calls "hyperobjects." Both Morton's and Barad's projects overlap with many of the conclusions I present here, but each manifests a kind of allergy toward the notion of Wholeness. They thus serve as exemplary ways of reasserting the language of fragmentation even when pointing to interconnection. While Barad doesn't insist objects are separate, she ignores the potential holistic aspects of her articulation, covering over the underlying unity with the ambiguous notion of the apparatus, and conflating epistemology and ontology. While Morton separates ontology and epistemology, he does so by insisting that objects are separate and thereby denying these aspects of Wholeness even as he points to interconnection and scalar objects.

Barad's "agential realism" uses her detailed knowledge of QT and Niels Bohr to generalize Bohr's complementarity of apparatus and object—the notion that attributes of an object are tied to the apparatus used to observe it.[38] While Bohr limits complementarity to the epistemological situation of the laboratory, Barad generalizes the notion of the apparatus to any material configuration that "enacts a cut" by encountering reality (142). This generalized apparatus allows her to extend an epistemological observation into what she calls an onto-epistemological position: apparatuses don't merely come to know but performatively co-constitute the things they are coming to know (185).

My impression is that much of Barad's account could square with Bohm's but that some important differences arise, specifically in the attempt to avoid appearing to take a mystical or holistic position.[39] Thus, in a footnote on Bohm, Barad states:

> There is an important difference between Bohrian "holism" and Bohmian holism. Bohm's theory involves a radical holism; everything matters (owing to the radical nonlocality), like traditional holism. By contrast, for Bohr, if one wants to apply the term "holism" at all . . . it must be understood that holism is about (specific) differences (and specific connectivities) that matter—differences within oneness, rather than oneness as a seamless, all-encompassing whole. (459)

The difference here can be understood in terms of how one defines a "system." When Bohr speaks of a "whole system" he means all aspects of relevance for determining the attributes of a subatomic object. Bohr aimed to bring together the object and the observing apparatus, which was traditionally separate within classical physics. Thus, Bohr emphasizes the unity between the apparatus and object as one undivided system. For Bohm, however, Bohr's "whole system" is not a whole at all but rather a subset of the Whole that has been delimited within the epistemological situation of the laboratory. In reality, attributes of a subatomic particle exceed this limited situation and may have thresholds of relevance—that is, attributes such as position—before the observation.[40] When Bohm separates these epistemological issues from the ontological ones, he avoids the problematic and ambiguous notion of the "observer," which only approaches being unambiguous in controlled laboratory situations. For Bohm, the significant implications of QT—including the nonseparation of any "apparatus" and object—arise in how these "attributes" of subatomic particles require reference to that which exceeds them.[41] Thus, Bohm's articulation is also about "differences within oneness" but in a more complete sense—as the many differences that might affect any "object," including but not only those in a definable apparatus. Thus, the difference between Bohr's and Bohm's holisms would be better articulated as epistemological holism (Bohr), a focus on the unity between what is involved in the act of knowing; and ontological holism (Bohm), the unity of all objects which express differently in an ontological sense depending on their configuration within the whole.

When Barad generalizes Bohr's philosophy she critiques Bohr's focus on epistemology as inherently anthropocentric and suggests that he "mistakes the apparatus for a mere laboratory setup."[42] However, Bohr's notion of the "apparatus" as a definable entity relies on this limited setup, tied to epistemological questions. Otherwise, as in Bohm's account, one must speak about the Whole as the system out of which all potential differences might come to bear on any given "object," subatomic or otherwise. In other words, to make this move to ontology, the system must be expanded indefinitely. Yet, Barad moves Bohr's epistemological statements into ontology, from the lab to all interactions, while avoiding Bohm's more "radical holism." Doing so creates three major problems. First, it retains this confusion about knowing and being, which gives too much power to appearances and attributes arising from an epistemological process. This confusion stems from this expanded notion of the apparatus, which, second, only increases the ambiguity of the "observer." Finally, these two attributes allow her to simultaneously affirm and deny the existence of boundaries between "things." Take, for example, the following passage:

> Bohr argues against the Cartesian presupposition that there is an inherent boundary between observer and observed, knower and known. That boundary is differently articulated depending on the specific configuration of the apparatus and its corresponding embodiment of particular concepts to the exclusion of others. That is, the object and the agencies of observation are co-constituted through the enactment of a cut that depends on the specific embodiment of particular human concepts. (154)

Like Bohr, Barad is suggesting that at the quantum scale one cannot look at objects as inherently separate from the observer. But while Bohm's ontological interpretation would drop this "from the observer" and say that "one cannot look at objects as inherently separate," for Barad, the term "apparatus" preserves this ambiguous entity. This leads Barad to develop a language that indulges the residual habit of fragmentation as these agencies actually "enact a cut."[43] In trying to emphasize the ontological nature of these apparatuses, Barad then situates them as material or "physical arrangements":

> Apparatuses are not Kantian conceptual frameworks; they are physical arrangements. And phenomena do not refer merely to perception of the human mind; rather, phenomena are real physical entities or beings (though not fixed and separately delineated things). Hence I conclude that Bohr's framework is consistent with a particular notion of realism, which is not parasitic on subject-object, culture-nature, and word-world distinctions. (129)

In arguing that apparatuses are inherently physical entities, Barad emphasizes separation, isolation, and cuts in the very argument about nonseparation and entanglement. Thus, while she argues at one point that "separability is not inherent or absolute, but intra-actively enacted relative to specific phenomena" (339), she nonetheless retains a different, agential yet ontological separability: "What replaces . . . spatial separability as the ontological condition for objectivity is agential separability—an agentially enacted ontological separability within phenomena" (175). Thus, in Barad's extension of Bohr's language, the apparatus is treated as a site of ontological fragmentation. The question will arise: how do you cut the apparatus from the rest of being? By another apparatus? And we have again Zeno's regress unless we follow Bohm's answer: even the apparatus does not exist independently from the Whole. But this final move is the one Barad avoids in dismissing Bohm as a "radical holism."

The difficulty with Barad's apparatus is mirrored in Timothy Morton's relationship with objects. Morton's work is situated within philosophies that explicitly focus on objects, treating objects as "real entities whose primordial reality is withdrawn from humans."[44] This approach inverts the problem just identified in Barad: what determines what an object is? How do we handle the fact that, in scaling, the same object becomes another object? What Morton calls "hyperobjects" are essentially scalar objects; they are "things that are massively distributed in time and space relative to humans" (1). In general, his descriptions of hyperobjects provides an insightful phenomenology of scale. Yet because of his philosophy of objects, Morton is unable to clarify the bewildering nature of these objects. While he says that "we are inside them," the commitment to the preexistence of objects leads to a confusing and rather dark articulation (20). He argues that "what the ecological thought must do, then, is unground the human by forcing it back onto the ground, which is to say, standing on a gigantic object called *Earth* inside a gigantic entity called *biosphere*" (18). And yet, in humiliating this human, he positions the situation as the inaccessibility of a spectral entity: "The gaps and ruptures are simply the *invisible presence* of the hyperobject itself, which looms around us constantly" (76).

I'd argue that we have here again a case of an epistemological limitation being confused with an ontological inaccessibility: just because you can't know something doesn't mean that this thing preexists in itself as a separate, forever-withdrawing object. Interestingly, this confusion arises in Morton's interpretation of Bohm. He is rightly excited that Bohm's ontological interpretation "takes Bohr's 'indivisibility' to pertain to objects beyond (human cognition)" (43). But, true to the allergy toward Wholeness, Morton immediately suggests that "according to the Bohmian view, you aren't part of a larger whole. Everything is enfolded in everything as 'flowing movement'" (14). It is unclear how Morton can suppose that Bohm holds this position given everything Bohm states explicitly in *Wholeness and the Implicate Order*. It seems to me that you can't, as Morton would have it, have your interconnection and fragment it too—unless you scale. And when you scale, you acknowledge that difference, the very grounds used to designate these objects as things to be in themselves, is not set in place in one way in advance. You can't designate this vast interconnection a "mesh," declare that this mesh arises from a viewing from a larger perspective, *and* deny the Wholeness aspects: "It's strictly impossible to equate this total interconnectedness . . . with something beyond us or larger than us. Total interconnectedness isn't holistic."[45] But what is this total interconnectedness other than viewing things according to their function within the Whole?

We are not only inside hyperobjects, *we are those objects already;* and everything we know and feel and are already is, in certain ways, a part of those objects. The affect here changes significantly. This is the whole gist of the turn to mysticism: there is quite a different affect associated with the great Upanishadic mantra, *tat tvam asi*—you are that!—in contrast with Morton's "claustrophobic horror of actually being inside it."[46] Both connect the vast to the here—*all of this is also Brahman, the Whole*—but one recoils in horror from this configuration while the other embraces it and reworks one sense of self around this new perspective.

ONE/MANY, TRUE/FALSE: REDISCOVERING THE DIFFERENCE IN PLATO'S *SOPHIST*

And yet, there is an understandable and intuitive reaction to these statements of Wholeness or Oneness. As one recent philosopher put it, "For if being is one, then one must posit that what is not one, the multiple, *is not.* But this is unacceptable for thought, because what is presented is multiple and one cannot see how there could be an access to being outside all presentation."[47] The concreteness of the world beckons! Look at all these objects! If everything is indeed One, then how do we find ourselves with the many? Thus, scale's counterintuitive nature arises again in this push beyond a nonscalar objects-based lifeworld that we so concretely interact with. In this variant of the attempt to kick a stone as refutation, scale offers a provocation: ah, Dr. Johnson, do you kick a pile of atoms, the Earth, or the Cosmos itself?

Provocations aside, clearly the basic question must be reiterated: even if we accept the One-Whole-All as a legitimate articulation, how do we arrive at this apparent many? Ogling the multifarious scalar forms, we can wonder: how is this multiple not just one multiple but many different multiples at once depending on how we look at it? On what grounds are scale domains produced and transformed again across scalar thresholds such that this concrete world manifests in multiple multiples?

It is worth returning to Plato's articulation, since Plato is often cited for his attempts to move from Parmenides's One to allow some notion of the many. For example, Lovejoy's classic *Great Chain of Being* suggests that Plato's move from the One to the Many is a contradiction that he calls the "principle of plentitude": that, in order for the All to really be all, it must also include the Many.[48] Rather than examining the simple yet profound logic of this statement, Lovejoy declares that Plato took literally a naive statement from "common speech . . . that it takes all kinds to make the world."[49] Lovejoy argues that Plato ultimately provides a "combination [that] may seem to many modern ears unconvincing and essentially verbal, and its outcome no better than a contradiction."[50] However, Plato's argument in the *Timaeus* fits with our account of the All in 3.38. Plato's formula "that the Whole may be All" (41c) uses for "All" the word *hapas,* an intensification of *pan,* variously translated as "absolute all" or "whole." If Timaeus is going to describe this universe, the ordered All, then he must not leave anything out. If anything is left out then it is not *hapas,* the Whole-All.

While Lovejoy examines the *Timaeus,* Plato's *Sophist* couches the movement from the One to the Many in the terms introduced already by Dick and Hussey: what is *to on,* "that which is"? Usefully, the *Sophist* situates this discussion within the dialectic between the idealist and materialist positions (see 246a-c) and the questions about Being and Becoming following Parmenides.[51] At the same time, the *Sophist* clarifies the stakes of this philosophical discussion by positioning it within an attempt to define the infamous sophist—the name given for the controversial teachers proclaiming to teach wisdom, rhetorical prowess, and statesmanship. In trying to define the sophist, Plato notes that a sophist might use Parmenides's argument to defend his sophistry: if we can never speak of "what is not," as Parmenides argues, then how could anyone claim that what sophists say is not true and that they are not truly purveyors of wisdom as they profess to be? After all, what they say, properly speaking, is. "It's extremely hard," says the visitor who has come to discover the sophist with Theaetetus, "to say what form of speech we should use to say that there really is such a thing as false saying or believing" (236e). Surely, to say that anything is an appearance, an illusion, or in any way false, one must have a clear idea of what kind of being a false thing is and how it relates to this *to on*.

Plato's solution to this problem of appearance is *not* to deny the One-Whole-All. Despite a joke about committing patricide of their "father" Parmenides, Plato's interlocutors follow Parmenides's separation of Form I and Form II, albeit with a particular account of how one moves from "that which is" to "that which is not," that is, that which only "seems to be" or is false. His solution is simple: "that which is not" (*to me on*) is not a negation of "that which is" (*to on*) but rather is a *particular form* of "that which is." The essential maneuver is to examine the nature of difference: if one can admit, to any degree, difference to be a part of "what is," then you find yourself with a role for "that which is not."

Plato's major departure from Parmenides is that he allows motion and change into the cosmos. But to do so, he only needs his interlocutors to acknowledge what we acknowledged at 4.2: that "that which is" is not entirely without grounds for differentiation within itself, whether we call this "fluctuation," "motion," "energy," or something else. In this case, Plato's evidence that *to on* is not absolutely immovable is the existence of *psukhē* (mind or soul), *nöos* (mind or understanding), and *zōēn* (life) (249a). In other words, because anything appears at all and things appear to change or grow, *some* kind of grounds for differentiation is part of what is. But how does this differentiation relate to this One?

As we noted in 4.1–3 about fluctuation, even motion and rest imply "the different" and "the same" (255a). Ultimately, the different has to share with all of these possible categories ("what is"; "what is not"; motion; rest; difference), since difference is required to differentiate them. Here, the different is not a thing but a type—an *idea:* "we're going to say that [the different] pervades all of them, since each of them is different from the others, not because of its own nature but because of sharing in the type [*idea*] of the different" (255e). Here is the crucial distinction: things differentiate not because they are different by their own nature but because one can apply the possibility of differentiation to them. This difference introduces a "that which is not" *within* "that which is" precisely because it distinguishes it out of the Whole: "So it has to be possible for *that which is not* to be. . . . That's because . . . the nature of *the different* makes each of them not be by making it different from that which is. And we're going to be right if we say that all of them *are not* in the same way. And on the other hand, we're also going to be right if we call them beings, because they have a share in that which is" (256b). Such differentiation positions one part of what is in relation to another part of what is, marking off certain things and providing them with names as a "sort of setting of being over against a being" (257e). In a more general sense, "since we showed that the nature of *the different* is, chopped up among all beings in relation to each other, we dared to say *that which is not* really is just this, namely, each part of the nature of the different that's set over against *that which is*" (258d-e). Is this not already what scale changes demonstrate: that reality finds itself distinguishable in different ways whenever any portion of itself encounters another portion? The *to me on* (not-being) does not enter into being itself qua *hapas* (Whole-All), but in the way that parts of reality are put up against each other.

Plato gives four sites where this setting of being against being occurs: first, at 260e, speech (*logos*), belief (*doxa*), and appearance (*phantasia*)—to which he, at 263e, adds thought (*dianoia*). Each of these is a *site for differentiation,* where a part of what is interacts with another part of what is to produce a difference that can make a difference. Plato has the visitor focus on speech, noting that its relational structure is clear—"it has to be about something" (262e). He argues that the others "are all the same kind of thing as speech" (264b), since each is a way that "that which is" sets itself against itself, permitting a kind of marking off or a chopping into bits via differentiation.[52] Here, I would suggest that Plato is recognizing a problem of logical typing (see 3.6), or what Ryle calls a "category mistake."[53] We provide a name—"that which is"—and then assume that "that which is not" is inherently its opposite. But what if "that which is not"—the true/false dichotomy as we usually conceive of it—is not the same kind of thing as the One-Whole-All of Being itself? The problem is that *logos* (whether translated as "speech," "accounting," or "reason") is *about* something else other than itself but which is still part of the same Being; it is one part of reality put in relation to another part of reality. This is to say that speech is semiotic in character: it picks some "thing," for example, a sound wave or a mark on paper, and that thing is put in reference to another thing.

The essential leap is that appearance too is grouped together with language, thought, and belief. This is the more significant move that subsequent philosophy helps us make sense of: when Kant asserts that there are a priori categories of reason that construct experience, he is pointing to how the differentiating process of perception requires certain selections of qualia and forms. In turn, when Charles Sanders Peirce first outlines his theory of signs, he conceptualizes this theory as a new set of categories in this Kantian sense.[54] Peirce's Kantian yet semiotic categories bring our conception of appearance in line with our conception of language in just the way that Plato was already experimenting with in the *Sophist*. That is, even perception or appearance takes on a semiotic character; for example, a light pattern is taken by the brain to be grounds for differentiation for the production of experience. In this account, differentiation is already a logical type above Being (that which is) because it is about Being—it is a selection of differences that come together to produce perception by setting a portion of Being against Being.

This setting of a part of Being against another part generates the Many as a realm of differentiation and provides means whereby one might assert truth and falsity within the Many. This account does not preclude forms of *epistēmē* (practical knowledge) but rather points to how we need to augment such knowledge with a systematic accounting of that differentiation. Scale is one way we trace out and make consistent the differentiations made in perception, thought, belief, and language.[55] Modern science attempts to more thoroughly prod the possible ways of differentiating reality using new technologies of perception and experiments designed to separate and identify possible grounds for differentiation. These tracings are necessary for combating the kind of scrambling of differences exemplified by the Sophists, which we see today in "post-truth" public discourse. But science is nonetheless a form of knowledge that deals with the Many; its truth is about and within the possibilities and layers of differentiation. If we assume otherwise, we find ourselves with a problem: do we then acknowledge that even these perceptions of the world require a selection of the mode of differentiation? Will we be open to this persistent re-differentiation? The key element here in recognizing this differentiating process is *aporia*, the sense of not-knowing: to recognize this configuration, the interlocutors must first work themselves into an uncertainty (*aporia*) about what exactly we might mean by these words "that which is" and "that which is not."[56] In this *aporia*, Plato points to another kind of knowing, not *epistēmē*, but *gnosis*: "Some imitators know [*oida*] what they're imitating and some don't. And what division is more important than the one between ignorance [*agnosis*] and knowledge [*gnosis*]?" (267c). The knowledge and ignorance of *gnosis* is about whether or not one knows the nature of these imitations and differentiations. *Gnosis* is the apprehension of the whole situation in which the different is drawn out from and within what is. Without *gnosis*, we risk assuming that the tracking of the possible differentiations within the Many (e.g., in science) is the acquiring of *gnosis* rather than *epistēmē*. As such, the truth-seeking of science and the truth-seeking of gnostic practices are fundamentally different operations.

It is in the context of *gnosis* that we find the rhetorical weight of the declaration that "all is an illusion." This mystical injunction is not of the same logical type as the operations of science but is rather an attempt to get one to observe this larger differentiating process whereby anything at all is able to be discerned. In the *Sophist*, Plato may avoid this larger point by suggesting that there are "things themselves" that are then copied naturally in dreams and appearances (266c).[57] However, articulating the situation in terms of copies and appearances points to an enduring trope of this more general maneuver as exemplified in the final lines of the *Diamond Sutra:*

> A shooting star, a clouding of the sight, a lamp,
> An illusion, a drop of dew, a bubble,
> A dream, a lightning's flash, a thunder cloud—
> This is the way one should see the conditioned.[58]

All of these phenomena are fleeting, deceptive, or somehow confused. What if these phenomena bring our attention to something more general about the differentiation process already being undertaken to generate any objects at all?

PONDERABILITY AND SUBSTANCE BEYOND THE HUMAN

What Dick experienced and what scale has brought us to is this simple fact: there are no objects, properly speaking "in themselves." However, Bohm's ontological argument, Dick's experience, and our zeroing out between scales all point to a reality "in itself"—what Kant calls the noumenon—that is not exclusive or reliant on one-who-experiences. However, this noumenon does not and cannot exist as anything less than the One-Whole-All. Kant confuses this point by speaking of an "in itself" for objects:

> I do not mean to say that these objects are a mere illusion. For in an appearance the objects, nay even the properties we ascribe to them, are always regarded as something actually given. Since, however, in the

> relation of the given object to the subject, such properties depend upon the mode of intuition of the subject, this object as *appearance* is to be distinguished from itself as object *in itself*.[59]

To avoid this possibility of objects being illusions, Kant relies on the fact that objects are "regarded as" given. Doing so leads him to distinguish between appearance and the "in itself" as both applied to presupposed objects. In turn, Kant banishes any access to this "in itself," thereby opening the possibility that objects exist in themselves while nonetheless remaining inaccessible.

To untangle this confusion we should again be wary of blending epistemological limits and ontological foundations. Following Bohm and Plato, we can say that the grounds of differentiation can exist in what is, *to on*. The perception of such grounds, however, requires situating one part of reality with another. This process creates a subject-object configuration. It is only if we preserve any "in themselves" for either subject or object that we would assume that the illusory and dependent nature of objects means that they are somehow created by the subject. Quantum physicists are not conjuring up quarks with no basis outside their brains (a possibility that assumes a great deal about subjectivity and human beings); they have developed complex apparatuses for identifying a difference discernible in Being under certain conditions. However, neither the apparatus nor the object exists "in itself." Again, we can invoke the essential scalar point: if a different set of differences are selected, atoms will appear—or cells or bodies. Do these differences belong to a pre-given object? If so, which one—the quark, atom, cell, or body? Properly speaking, then, the "in itself" as grounds of differentiation belongs to the Whole, not as Whole (*hapas*), but as differences that are, in Bohm's terms, nonlocal properties discerned within it through an implicate ordering.

Therefore, the "in itself" will never appear as an object. But how then will it appear? Even though Kant denies it is possible, Dick nonetheless experienced the constructed nature of these categories. Dick understood this point quite well: "Perhaps you see Kosmos—continuum—when the Kantian categories go because space, like time and causation, is one of them; . . . I'm saying that all three go: time, certainly; that most of all. . . . And also space: discrete plurality gives way to unity because 'space' equals 'void' and is a Kantian category, which is to say 'discontinuous matter' is connected with these categories and is false" (55:1). But, logically, if the categories fall away then so does experience. How could one experience the falling away of experience?

By examining the process of observation itself. Here we can turn to the American philosopher Franklin Merrell-Wolff, who was a careful reader and practitioner of not only Advaita Vedanta and Buddhism but also Kant.[60] When describing his "ineffable transition," he speaks of "abstract[ing] the subjective moment . . . from the totality of the objective consciousness manifold."[61] In other words, he was able to separate awareness itself from any content or object produced for it. Merrell-Wolff analyzes this abstracting process in relation to Kant's categories, arguing that this was, properly speaking, *not* an experience in the Kantian sense, since any experience requires a priori categories. In such an encounter, Merrell-Wolff argues, subject and object drop away, but not merely as a blacking out of awareness but as pure awareness without content. Thus, he calls this state "consciousness-without-an-object."[62] While we usually think of consciousness as always "consciousness of," this presumes that the attribute of awareness can only be ascribed to objects within consciousness.[63] But what if one becomes aware of the fact of consciousness itself apart from any object that can serve as content for it?[64]

The result is a kind of zeroing out best expressed by the Buddhist term *śūnyata,* emptiness.[65] One might think that *śūnyata* is contrary to our discussion above about *to on*. However, Nāgārjuna's defense of *śūnyata* in *The Middle Way* articulates something quite similar. Consider the following verse: "If something that is non-empty (*aśūnyam*) existed, then something that is empty (*śūnyam*) might also exist. Nothing whatsoever exists that is non-empty; then how will the empty come to exist?"[66] This argument is a mirrored version of the argument about Being. *Śūnyam* implies that a thing does not have an "in itself" because it lacks stability and is dependent on something else both for its origination and its perception. Thus, Nāgārjuna states "Dependent origination we declare to be emptiness."[67] Similarly, Nāgārjuna uses the term *svabhāva*—literally, "one's own being"—to argue the same: "But how could there ever be an intrinsic nature [*svabhāva*] that is a product? For intrinsic nature is not adventitious, nor is it dependent on something else."[68] If we admit that anything is empty in this way, then we have to ask: what thing is not empty? Is there any "thing" that doesn't pass away or require something else for its origination or observation?

Thus, consciousness-without-an-object is a not-knowing and a non-experience in the Kantian sense that is best expressed as emptiness.[69]

Crucially, *śūnyata* does not itself exist as an attribute among other attributes of things. Rather, *śūnyata* is another rhetorical device for orienting us to our situation, focused on the habitual way we treat objects as real. If we treat a term like *śūnyata* as an idea or category of reason, like Plato's difference or Kant's a priori, we miss the point: "Emptiness is taught by the conquerors as the expedient to get rid of all views. But those for whom emptiness is a view have been called incurable."[70] Instead, the entire operation is a negation of *any* object of consciousness so that one moves from the Many to the Whole, from any content of *hapas* to a complete zeroing out of the sense that any thing has intrinsic reality.

The result is that one comes to see objects as a hypostatization. *Hypostasis* means "setting under." In Plotinus the term is used for those ways that Being manifests that are grounded within this Being—the One, the Nous (intellect), and the Soul.[71] The term can be used in this positive sense for ways in which things are undergirded by substance—as when Dick says that "since creation is a hypostasis of God . . . one should look for beauty in it, as manifestations of the divine" (91:J-89; 755). But this definition leads to a negative meaning: that these things are *not* the substance, and hypostatization is the attempt to grant substantial status to what lacks intrinsic nature. Thus, in an earlier passage, Dick says that

> in 3-74 I stopped hypostatizing, and there lay the divine stuff *in place of* everything. The *they* were gone: the *it* shone through. From what had been high, low, me, not me, small, large, important, trivial. Liberation, not theoretical but actual. This is the universe organism which I perceived to be the macrocosm—alive and unitary. (24:1; 217)

Here, hypostatization means to consider things as real, to ascribe them an "in themselves." Siderits and Katsura use hypostatization in this way in translating the following Nāgārjuna verse: "Liberation is attained through the destruction of actions and defilements; actions and defilements arise because of falsifying conceptualizations; Those arise from hypostatization; but hypostatization is extinguished in emptiness."[72] Here, actions (*karma*) and defilements (*kleśa,* variously distress or affliction) arise through *vikalpa,* a distinction or difference in perception that, in texts such as Nāgārjuna's, also comes to mean falsity. Hypostatization is a translation of *prapañca,* the reification of the diversity of phenomena or excessive (investment in) wordiness.[73] In other words, distress arises from taking one's distinctions as reality. This, in turn, is extinguished in *śūnyata.*

Merrell-Wolff sums up this situation with the formula "substantiality is inversely proportional to ponderability."[74] "Substance" here means "the substrate underlying all experience, which is not itself a direct object of experience" precisely because experience must select differences out of which to render experience.[75] The things most concretely before us—for example, these books, chairs, foods, other people—are those most saturated with differential forms and weight and therefore most removed from substance. To rewrite this field of awareness in terms of cells or galactic events, we have to decrease this ponderability (what we described at 1.27 and 2.9 as a zeroing out). Yet this scalar operation creates new kinds of ponderability. Thus, even though these scalar objects are less ponderable than the immediate sense-world, they are still not substance. At the same time, Merrell-Wolff argues that "there is nothing in this to invalidate the positive findings of natural science," only that it reorients our relationship to them.[76] This view suggests that "the objective world . . . is a precipitate from Consciousness in its most comprehensive sense, but it is *only partly determined* by perceptual and conceptual consciousness."[77] The very partiality of the perception produces the sense of an outside, since only a portion of what is is brought into the perception and conception. In actuality, this independence ought to be conceived "as negation instead of positive actualities." Only then, suggests Merrell-Wolff, can we "proceed with the systematic development of either a science or a philosophy."[78] In turn, "none of this implies that experience is wholly without value but its value is symbolic and instrumental." Science can proceed for the purposes of *epistēmē,* as long as we understand its relation to this other form of knowledge and truth (more on science in chapter 10).

In contrast, *gnosis* is what Dick intuited through a crucial kind of surrender: the surrender of hypostatization or the habit of division itself.[79] Thus, in the regressing object-world of *Ubik,* Joe Chip only finds Ubik when he gives up, sitting on a park bench and releasing "some of his vast inertial weight."[80] Similarly, Grote, the philosopher trapped in the Zeno's paradox chamber, must give up before he can escape: "how long would it be? . . . Suddenly a terror rushed through him. 'Maybe I shouldn't figure it out.'"[81] In turn, Dick experiences over and over again these moments where he, too, gives up—but not for despair but toward this larger *gnosis:* "I give up. Its hold was broken over me in 3-74—Salvation is real" (21:17; 416).

TOWARD AN INFORMATIONAL ONTOLOGY

How, then, are we to describe objects?

Dick will throw into the mix the final provocation of this chapter: the assertion, first published in the novel *VALIS,* that "The universe is information."[82] Not only this, but objects are specifically hypostatizations of this substance now called information: "We have hypostatized information into objects."[83] What does this mean?

We can start from the point discovered in 1.19: that objects are those differences that are able to make a difference at a particular range of observation. If we combine this with Bateson's definition of information as a "difference that makes a difference," then we seem to be implying that objects are, properly speaking, information. Why has scale led us here, and what does it imply? Let's consider more closely Bateson's definition:

> Kant argued long ago that this piece of chalk contains a million potential facts (*Tatsachen*) . . . but that only a very few of these become truly facts by affecting the behavior of entities capable of responding to facts. For Kant's *Tatsachen,* I would substitute *differences* and point out that the number of *potential* differences in this chalk is infinite but that very few of them become *effective* differences (i.e. items of information) in the mental process of any larger entity. *Information* consists of differences that make a difference.[84]

Bateson uses the example of running one's hand over a surface to identify a raised surface: information is provided by the signal of differences detected by the hand. He suggests that all senses work in similar ways, using different media to detect differences.[85] Later, Bohm adds to Bateson's definition: "Information is a difference of form that makes a difference of content, i.e., meaning."[86] But by "content" or "meaning" here, Bohm means the effect that a form will have—how it will "in-form" in the more literal sense of imparting form. He points, for example, to a radio wave: "The form in the radio wave thus literally 'informs' the energy in the receiver, i.e. puts its form into this energy, and this form is eventually transformed (which means 'form carried across') into related forms of sound and light."[87] Both Bohm and Bateson suggest that this operation is far from an exceptional kind of thing, but exists anywhere we find patterning and the conveyance of structures of form. Thus, as further examples, Bohm discusses the aggregate behavior of electrons and Bateson discusses the repeated patterning of spirals in biology.[88]

This imparting of form might be read in relation to epistemology, that is, considering how differences function for a structure of mind. However, the observations about difference at 1.19 and 3.28 and Bohm's ontological interpretation suggest that such differences might make a difference in a broader sense, with epistemology simply a special case of this information flow.[89] Objects become information in the sense that they are whatever can make a difference at a particular scale; they are what can pass along form as patterns of difference. In Bohm's formulation, just as radio waves are passive (yet present) information that is only made active by the encounter with a receiver, there is passive information in any system that is made into active information by the energy of an object such as an electron.[90] In scalar terms, we might say that a water molecule does not make a difference at the scale of a meter because its form is too small to be transferred; it must come together into a larger form—the water in a glass, perhaps—to be a form that in-forms (or is in-formed by) the glass (see 6.18–19). Likewise, the human body shapes the formal arrangement of its aggregate cells and molecules, but it does not in-form the individual cells and molecules; this occurs in the interaction and structure where the whole "form" of the body is no longer directly available as a mode of differentiation. Undoubtedly, cross-scalar "information" exists, but it would have to be a product of pattern, which will always mean an aggregation of differences or a constraining of possibilities of differentiation (see 6.30–36).

This conception of objects may seem to grant a kind of agency (as Barad does) or in itself (as Morton does) to these objects. We must tread carefully here: the passing of form *is* the aspect that makes an object appear. Does this ability to in-form belong to a preexisting object? To the contrary, we can suggest that objects are *signs* of patterns of existence as they aggregate. Thus, second-order cyberneticist Heinz von Foerster defines objects as constraints arising due to eigenvalues. An "eigenvalue" or "eigenbehavior" is a stability that arises from combined or recursive values in vector mathematics. Foerster extends this mathematical term to any stabilities that arise from combined behavior.[91] Using 4.12, we could say that reality presents itself with emergent stabilities at a particular scale that are a result of activity of lower scales. An object exists in a seemingly external sense through this stability. Attributes then become linked with—further stabilized onto—an object as a sign of the capacity to work with particular differences (i.e., constraints) localized onto that object. Thus, Foerster describes "objects as signs, symbols, 'tokens' for eigenbehaviors."[92] In this account, objects are information when portions of reality are gathered together and able to function as a unit.[93] Using this eigenbehavior, an object becomes a site for meaning when encountered by another unit capable of making this information active.[94] In the case of humans, the various patterns of difference in a sensory field combine with all of the patterns of difference of the body on multiple scales (bodily needs, social expectations, ecological pressure) to identify an object (e.g., an apple) as something of note and of a certain character and value. This is, in a sense, to entangle epistemology and ontology, but only insofar as one must have parameters for differences to make a difference. Once these parameters are selected, they are part of the ontological situation, but in a way that must also be distinguished carefully through a tracing of the epistemological configuration. This is what scale performs: it marks an epistemological configuration required for an ontological presentation so that we can say with some clarity what arises from what part of the configuration.

Dick notes that "our problem in distinguishing the structure stems from the fact that much of this information . . . *is not recognized by us as information at all*" (55:41). Once we see this information aspect, we are able to view these objects as parts of a larger structure, arranged together in a form that arises from their role in the larger structure: "its structural basis—that is, it constitutes a unity—on the basis of information" (53:14). But again, Dick is not implying that objects are just what you make of them. If they were, then no description on other scales would be possible, since scale requires the rewriting of objects in your lifeworld. Instead, there is a disinhibition of the usual stance toward objects so that we are able to see these objects in relation to the total system:

> What I saw about the external disinhibiting structure which evidently surrounds each human being . . . was the utilization of every sort of datum, especially visual, so that when required that particular datum projected a signal . . . which the intended person to be disinhibited received. . . . The intended individual would experience a sudden transformation of the ground-set formation of the environment around him;

> one item would come forward, alter from ground and become set, then go back once more, to resume its passive or inert mode. (4:147; 70)

Observing our relations to objects is a way of tapping into and observing the structure of possible relevance encoded into any object. After all, why do certain objects become relevant? Or, even more, why does scale make new objects relevant in ways that they weren't before? Consider, for instance, how "climate change" must conjure up a set of relations that reworks certain objects (your automobile) in relation to certain other objects (the hurricane hurtling toward the coast). Suddenly, we experience a "transformation of the ground-set formation of the environment" around us that takes the form of a cross-scalar relation. Such relevance requires a disinhibition to provide new room for this system of linking. Looking from one object (this hunk of metal that burns gas to move me around) to another (a chart showing carbon measurements), a new linkage and significance can be forged.

GROUND-SET DISORDER AND DISCOVERING THE CONDITIONS OF PERCEPTION

But how do we arrive at this view of objects? Dick notes that "the information basis of reality became available to me, due to the meta-abstraction" (53:39). In turn, this notion of information helps us understand the meta-abstraction process performed by scale.

Bateson notes that "it takes at least two somethings to create a difference."[95] He argues that overlaying two perspectives forms a logical typing, which produces new information about that information. For example, in order to produce binocular vision "the difference between the information provided by the one retina and that provided by the other is itself information of a different logical type. From this new sort of information, the seer adds an extra dimension to seeing."[96] When we arrive at a new scale, just such an operation occurs. To consider the Cosmos as atoms or galaxies is to provide a second mode of vision, a whole new set of differences, that provides a new dimension to reality (see 2.5). This overlaying does not just provide multiple ways of dividing the Cosmos; it also, as we have shown throughout this chapter, provides a reflection on the very dividing process itself. Thus, while we undoubtedly can flesh out objects and relations at new scales, this new information about this differentiating process is always available as soon as we overlay any two of these perspectives. Hence, the basic provocation of this book functions by overlaying these perspectives within a single object: how is it possible that this body is both cells, galaxies, atoms, and a part of an ecology?

Here we have Dick's "quantum leap in understanding," which "involves a meta-abstraction" (55:41). There is a moment, in thinking about these layers of being, where we realize—

> Wait a minute: I'm missing something *terribly* important. In 2-74 I realized about the two coaxial worlds here sharing one space-time. In 3-74 I saw the set-ground; I then was *discriminating the two worlds each from the other*—and yet this was only an extension into particulars of the 2-74 meta-abstraction itself! The meta-abstraction was in itself a *seeing* of this; once the concept was there the rest did follow. (55:41)[97]

In binocular vision, this process of patching together these perceptions to create depth has been integrated into our perceptual-conceptual system. With scale, however, we find something new—an extension beyond this system of perception that is about this system. It provides new depth, but a depth that has to be integrated, properly understood, and adequately assimilated into our approach to the Cosmos. Hence, the radical reconfiguration that scale has launched us into.

NEGATION AND AFFIRMATION IN THE MICROCOSM

Dick makes this reorientation inherently personal. In this register, the One and Many might be more adequately articulated as the microcosm-macrocosm or the Atman and the Brahman. These distinctions are directly about the reconciliation of the personal (here) with the infinite One-Whole-All. In the continuum model, the undoing of our habit of division rewrites subject-object interactions within a scalar relationship: "and of course in kosmos—truly understood—the percipient is part of the structure, not outside it, and ideally shares in its mind. . . . 'inner-outer' signify only 'macro-micro'" (53:6). By relaxing this habit of division in relation the object-subject called "me," objects themselves take on a different role and form: "a given object or event can play one role in our world [the discontinuous view] and quite another in *that* world [the continuum view]. In *that* world its role is one of pure function as part within one unitary integrated system: the part derives its identity, meaning, significance, and purpose *from* the total system, and, alone, signifies nothing at all" (53:14). Here, the *Homo sapiens* does not become the reference point within the Whole; rather, the Whole becomes the reference point of value for all objects, but *only if you integrate yourself into it* as part of the Whole.

In continuing to attend to his own scalar experience, Dick examines this mass of pages he has produced and declares:

> I *qua author am a function of it!* I am a mouth piece for it. . . . My corpus of writing is a true picture of the reality situation, since the macrobrain is the actual author. But the "audience" isn't us here but the *outside;* then Zebra, the macrobrain (Logos) is in here, inside this "bubble" with us. Reporting back out to its source. Of course it's here with us; I *saw* it; we're *in* it. (21:14–16; 416)

In making this statement, Dick is not asserting control but yielding it up: "Zebra continually guides (controls?) us as we move (are moved) through the 'maze' of life—disinhibited constantly and at the right time and place by the right signal" (25:11; 230). The negation of a negation, the selective undoing and remaking of divisions within the Cosmos becomes the very path of our way through the "maze of life."

Thus, the most clearly scalar vision in Dick ends with a reconfiguration of control. In *Divine Invasion,* this "Hermetic transform" travels through this inner-outer Möbius strip:

> By degrees, the transform took place. He saw outside him the pattern, the print, of his own brain; he was within a world made up of his brain, with living information carried here and there like little rivers of shining red that were alive. . . . Meanwhile he introjected the outer world so that he contained it within him. He now had the universe inside him and his own brain outside everywhere. His brain extended into the vast spaces, far larger than the universe had been. Therefore, he knew the extent of all things that were himself, and because he had incorporated the world, he knew it *and controlled it.*[98]

To access this configuration, the character must mirror within himself the pattern of living information both within and without his brain. Within this introjection, understanding comes from converting the inner-outer into the micro-macro. In doing so, he ends with this sense that, having incorporated the world, he knows it and controls it. But this knowledge and control is now entirely different: one knows and controls it *as the totality of that existence itself,* not *as the small-scale object in a* Homo sapiens *body.* In other words, one knows and controls it because one is it. Such "Knowledge through Identity," as Merrell-Wolff names it, emerges from a relaxing of the habit of division and the supposition that one (as body) controls or even is anything apart from the Whole.[99]

The control here is thus quite different from any human-based power, particularly the synecdoche that would claim the Whole for the human (see chapter 9). Rather than individual control, possession, or comprehension of objects, the end of a fragmented view permits us to work with those entities within the Whole. This, argues Bohm, is the ultimate result:

> It is proposed that the widespread and pervasive distinctions between people (race, nation, family, profession, etc., etc.), which are now preventing mankind from working together for the common good, and indeed, even for survival, have one of the key factors of their origin in a kind of thought that treats *things* as inherently divided, disconnected, and "broken up" into yet smaller constituent parts. . . . If he thinks of the totality as constituted of independent fragments, then that is how his mind will tend to operate, but if he can include everything coherently and harmoniously in an overall whole that is undivided, unbroken, and without a border . . . then his mind will tend to move in a similar way, and from this will flow an orderly action within the whole.[100]

As we hold these objects as real in-themselves, the forces of consumption and strife proliferate as we find ourselves struggling to hold on to these objects, acquire more, becoming jealous over what others have, and becoming confused by things that appear to be separate but are nonetheless intimately related. Dick's *Divine Invasion* dramatizes this contrast between fragmentation and wholeness as a battle between good and evil: the expression of Belial takes the form of the "accuser," the Jewish mystical name for Satan, which makes the characters see everything in their worst light. In a world of isolated objects, all regresses toward this dismal, fragmented sensibility. By contrast, Emmanuel, The Advocate, restores the sense of beauty to the world, allowing the individuals to appreciate what is already before them as a function of the Whole. Likewise, in our experience of scaling, these objects retreat, become small, are transformed but, most of all, are enmeshed with us in the larger Whole. Scale returns us there and requires this return if we are to permit this new configuration to truly in-form our sense of existence.

8

I Am the Transhuman Cosmos

Scalar Configurations of Subjects

THE VANTAGE POINT

Floating in the depths of space, Apollo astronaut Edgar Mitchell has a moment to pause and dwell on the view:

> Then I looked beyond the earth itself to the magnificence of the larger scene, there was a startling recognition that the nature of the universe was not as I had been taught. My understanding of the separate distinctness and relative independence of movement of those cosmic bodies was shattered.[1]

This scalar vision connects this *Homo sapiens* to a being far greater than the limited flesh and bone of this body. The sentiment echoes like an intergalactic Thoreau remixed with Carl Sagan: *The vast Cosmos! The actual universe! The common molecules! The exchange of starstuff! Contact! Contact! Who am I? Where am I?*

In processing this dislocation, this disorientation provides a significant and powerful new reorientation of Mitchell's whole way of seeing, thinking, and being:

> I experienced what has been described as the ecstasy of unity. I not only *saw* the connectedness, I *felt* it and experienced it sentiently. I was overwhelmed with the sensation of physically and mentally extending out into the cosmos. The restraints and boundaries of flesh and bone fell away. I realized that this was a biological response of my brain attempting to reorganize and give meaning to information about the wonderful and awesome processes that I was privileged to view from this vantage point.[2]

Significantly, the realization is not simply about the cosmos around him but about "Edgar Mitchell" himself as these "restraints and boundaries of flesh and bone" drop away into the scalar experience. In this statement, Mitchell launches us into the implications reached in 2.23–29: that scale radically reworks our notion of the "I." Scale reconfigures not just "subjects" but one's own sense of being-a-subject as a clear entity tied to a discernible structure. As scale presents experiences that exceed this seemingly apparent body, it renews the perennial question: what, then, am I?

Here, scale runs us into the much-heralded death of the humanist subject—the suggestion that the human as a clearly contained, autonomous, separate self does not hold up under closer scrutiny. Scale both provides one way of examining this new configuration and helps clarify the difficulty of this question of the subject. Who, we might ask, is this body called "Edgar Mitchell" floating in space? What of his situated viewpoint? What of the structure of vision as embodied, formed from an eye floating in space? Who, then, is experiencing this dissolution? After all, even in this transformation, this viewer notes "that I was privileged to view from this vantage point." In the current critical landscape, we move too quickly to the political and social constraints that appear to be and produce "Edgar Mitchell," reasserting this coherent subject at the very moment of its becoming-open, becoming posthuman. Undoubtedly, the white-male-scientist subject position was essential to getting this *Homo sapiens* "Edgar Mitchell" into space. Yet, even this position was not enough to withstand the overwhelming effect of this scalar vision. What is this becoming-vast such that Mitchell feels himself pulled apart and replaced by this sense of extending out into the cosmos?

Not only is the seeming paradox of this dissolution of the subject produced by scale, but scale provides some ways of renewing and reexamining the problems of subjectivity, identity, experience, and consciousness. Scale produces a situated dislocation (2.18); it dislocates the perspective from the delimited structure of the body even as it accounts for this dislocation. In scaling, one's identification with any object—especially the body—is dissolved to make way for the multitude of new objects (cells) and connections (ecological relations of the planet). However, experience remains localized in an interpretive configuration: seeing still occurs. Rather than reinscribe this experience as subject—as "me"—scale prompts us to ask again: who scales? To answer this question, we can attempt to find an "object"—the organism, perhaps—to designate as the one who is experiencing. Yet doing so leaves this subject untouched by scale. When we scale this object-made-subject, who remains to scale? Only one answer remains: this is the Cosmos, the Whole, scaling itself. The simplicity of the conclusion here—nothing more that Carl Sagan's "we are a way for the Cosmos to know itself" or the Upanishadic *tat tvam asi* (thou art that)—does not avoid the difficulty of this configuration. Despite the common declarations of the death of the self, the inessentiality of the ego, or identification with the Cosmos, we still find ourselves retreading this ground for good reason: how do "I" handle a configuration of reality that suggests that "I" am not what I appear to be? Even if one experiences this dissolution, the question remains: what is the relationship between that experience and this one?

These paradoxes of subjectivity are the counterpoint to the paradoxes of motion we discussed in the previous chapter. In this chapter I want to first highlight how these paradoxes of individuality have arisen in posthumanist philosophy and philosophy of mind. These will provide the grounds for tracing out specific maneuvers used to reassert individual subjectivity even in attempts to get beyond subject-object dualities—specifically, autopoiesis theory, critiques of disembodiment, critiques of totalizing perspectives, and calls for situated knowledge. While each of these in some way critiques nonscalar subjectivity, none provides a fully scalar configuration of the subject. Examining the residual individuality present in such critiques will open a space to more directly face the seemingly paradoxical nature of scalar subjectivity. I will situate the major conceptual difficulty in this distinction between the specificity of an experience and the attempt to separate out definitively that which produces the experience. In what manner is experience situated? This discussion is summed up by the scalar formula "epistemological specificity does not equal ontological separability." Finally, I will to return to the history of thought dealing with the nature of the human to suggest that this articulation might be called a "mystic transhumanism"—"transhuman" because it resituates the human within a cosmic context, and "mystic" because it transforms these subjects in the encounter beyond their supposedly contained identities.

DISLOCATING THE SUBJECT

In recent attempts to grapple with these changes in subjectivity, it has often been suggested that if we deconstruct the subject-object distinction, we are also getting beyond the various ways that humanism has conceptualized subjectivity. In early postmodern theory, this decentering of the subject was grounded in an argument that the ego or coherent self is an illusion.[3] As this critical stance has morphed into what is now called "posthumanism," this incoherency of the self seems to have been flooded out by a proliferation of theories that treat the subject as various enactments, performances, social regulations, and the like. The tentative or illusory nature of the self has become a more or less hidden assumption behind theories of posthuman subjectivity.[4] If the subject is able to be perpetually co-constituted then, by implication, the subject is somehow inessential. Our question about the subject who scales refocuses on this inessentiality and provides a new way to examine the contours of the subject and associated terms—mind, consciousness, identity, and individuality.

In posthumanism we see a shift in emphasis from deconstructing the subject to decentering the human. As an explicit scholarly current, posthumanism, according to Cary Wolfe, "names a historical moment . . . which [performs a] decentering of the human by its imbrication in technical, medical, informatics, and economic networks."[5] In this invocation of networks, Wolfe is pointing to a fundamentally scalar dislocation. The human is imbricated not just in the usual sociocultural networks of immediate this-scale experience but in networks that pull the human into other levels of interaction. In this entanglement, Wolfe argues that "the posthumanist form of meaning, reason, and communication" is "untether[ed] from its moorings in the individual, subjectivity, and consciousness."[6] In this statement he targets a notion of the subject as inherently present to itself as a clear, individual, autonomous agent of action capable of observing itself. The challenge from posthumanists is: are humans actually those things?

This critique of the subject is not exclusive to posthumanists; it also stands as a crucial moment in philosophy of mind, as exemplified by Gilbert Ryle's 1949 work, *The Concept of Mind,* which critiques the Cartesian myth in which the subject becomes a "ghost in the machine." Yet, Ryle highlights a difficulty: an essential part of his critique is that it is a mistake to think of mind and matter as two different kinds of things. In contrast, Ryle argues that the difference between mind and matter is an artifact of a logical typing—in his terms, a category mistake.[7] As a result, he becomes associated with behaviorism (a misapprehension anticipated by Ryle himself) and other materialistic theories of mind.[8] After all, if mind is not separate from matter, then it is easy to suggest that matter causes mind. One might ask the same thing of posthumanists: if the self, mind, and consciousness are inessential, then what is mind?

The various solutions to these questions first run into what Elizabeth Wilson once noted as the "compulsive antiessentialism" of critical theory: a denial that critiques of the subject are, in any way, reducing ourselves to matter or biology.[9] Wilson and others have since produced more nuanced readings, including the sort we already saw from Barad in the previous chapter. But we can wonder here if we've lost sight of why we've found ourselves in this tangle about the subject. Is this question not made newly fresh by pulling this "self" out of the apparent and intuitive fact of "me" as this body—or else, in the form Ryle critiques, this body possessed ("my body") by a presumed cogito? We then have this scalar choice again: do we seek another scalar object (now the cells or the social network or the atoms) to explain the presence of this experience, mind, and subjectivity?

One might notice that this question of the self or subject has already run us into the question of individuality: one seeks a subject in a form that is localizable as a separate, individual entity. Yet what if this intuition of individuality is an underlying scalism more fundamental than (but still inherent in) humanist subjectivity? This question about individuality even manifests in scientific discourse in questions about the descriptive accuracy and functional utility of asserting clear and inalienable individuality. For instance, in an article published in the *Quarterly Review of Biology*, Scott Gilbert, Jan Sapp, and Alfred Tauber argue against the functionality of individuality in biology. Picking up on posthumanist tropes, they argue that the discovery of the microscope and more recent technologies "lead us into directions that transcend the self/nonself, subject/object dichotomies that have characterized Western thought."[10] The article then surveys theories of individuality within biology and argues that each of these fails to take into account the essentially scalar relations of both the diversity of cellular structures and the microbes that saturate any given organism. This argument demonstrates how a presumption from this-scale experience (that "I" am an individual) confuses research on other-scale objects (e.g., the immune system response). Might we then ask how this presumption of individuality also corrupts and confuses our thought more generally?

In posthumanism this question of individuality is expressed by Wolfe's "we are not we" or "persons aren't persons."[11] In facing this statement we might nonetheless want to render as functional certain notions of individuality. Such attempts miss an essential difficulty: that our tendency to treat individuality as set in advance overshadows and continually corrupts our capacity to take into account the ways supposed individuals are not, in fact, individuals. Attempts at reinscribing individuality usually share a similar feature—an attempt to remain on or privilege one scale—and therefore avoid the most important reorientations inherent in scale. When, in preserving some sense of individuality, are we just searching for another thing to pin our sense of self on?

In facing this sense of self, a further question arises: are we speaking of the subject, individuality, and identity from the first person or the third person? This question was introduced by Thomas Nagel and developed by David Chalmers as a return to a bare phenomenology of consciousness—the "what is it like to be X?" In terms of scale, this question might be read as an attempt to return to the scale of here and ask again about the nature of experience. Nagel's original article begins by citing scalar shifts, asking what makes the "mind-body problem unique" in contrast to "the water-H_2O problem or the Turning machine–IBM machine problem or the lightning–electrical discharge problem or the gene-DNA problem or the oak tree–hydrocarbon problem."[12] In contrasting the mind-body problem to these clearly scalar examples, is Nagel denying that consciousness is also a scalar problem, or is he suggesting that we locate the problem at the scale of here? The former question can be found in Daniel Dennett's opening bewilderment in *Consciousness Explained*: "How on earth could my thoughts and feelings fit in the same world with the nerve cells and molecules that made up my brain?"[13] The latter is what Chalmers develops in his seemingly antiscalar argument: that consciousness cannot be described in terms of microphysical facts.[14] However, Chalmers's argument is also in scalar terms, beginning with an extensive description of the H_2O-water problem and then arguing that consciousness cannot be reduced to its components in this way.[15] In this form, one might argue that Chalmers is actually attempting to relocate experience at the scale at which it emerges: right here. In this form, Chalmers and Nagel highlight the possibility that the intuition of individuality and self is based on this specificity and irreducibility of experience. Perhaps we hold to individuality because on an intuitive level we feel that we are individuals, and on a conceptual level we need to account for the specificity of experience. After all, just as I am not currently experiencing Times Square in New York City, I don't experience other scales. Surely this particularity of perspective must be accounted for.

But still—is it mine? Who is that?

DELINEATING THE SUBJECT

Even after humanist notions of the subject are deconstructed, we can attempt to reclaim the self as an individual who is subject. Let's map a few options in terms of scale before turning to a few particular cases. The previous section implied four options, which in a scalar shorthand we can call up (network), down (physical/component), out (content "here" treated as objects that are "other"), and in (experiential qualia). Here we can consider these four directions in relation to three aspects available at this scale: the body (an object), the person (as a present social designation), and mentality (thinking, experiencing).

The body is the most readily available aspect: the intuition of myself as the body is based on this-scale experience, where this body appears as a distinct object that is mine. This is to read the body as "me" according to experiential qualia (in). One could also read the body in terms of the structure of the body that produces an "in": "I" am the body-apparatus that perceives and responds or is set apart from others via a particular location "here" (out). Alternatively, I can describe myself as a *Homo sapiens* who is differentiated according to evolutionary and ecological criteria (up). In a material reductionist mode we can describe the body according to its components, including selecting smaller-scale markers, such as DNA, to stand in for a sign of identification (down).[16]

One can also focus on selfhood in relation to social aspects present at this scale. One might treat the "I" as a political designation or socially reinforced reference point (up). We might note how this entity is created out of various material and economic conditions (down). Finally, politically and socially oriented phenomenologies treat the self as a subject who is a site for communal constraints and interests (in) or who, in being hailed by others (Althusser's interpellation), is constituted as the subject of address or action (out).

The target of posthumanism is usually the identification with the self as the mental apparatus. Yet even if we critique Descartes's *cogito*, Chalmers's argument could be used to suggest that "what I am" is the irreducible consciousness presented within experience (in). Alternatively, we can explain mental/inner life according to cultural or linguistic constraints (up), as the subject for the conditioning of behavior (out), or according to unconscious attributes, whether we follow the psychoanalytic notion of "the unconscious" or call these attributes the functions of neurons (down) (see Figure 8 for summary).

	Up	Down	In	Out
Body	Ecology	Molecular system	Apparatus that produces experience	Location (contra others)
Person	Political entity	Economic entity	Site of social significance for	Subject of address/action by others
Mind	Cultural/ linguistic subject	Unconscious or brain	*Cogito* or experiential specificity	Subject of conditioned responses

FIGURE 8: A topography of options for individuating the subject.

We can now examine some configurations of individuality, self, and subjectivity that could be used to critique accounts such as Mitchell's. These are essentially critiques of nonscalar views of subjectivity, but since they do not provide a scalar diagram of the subject, they have crucial ambiguities that need to be ironed out if we are to find a scalar configuration of subjects. Examining these will therefore highlight problems in our notion of subjects, identify ways scale resituates these problems, and introduce some essential aspects of a scalar notion of subjects.

LOSING THE MIND IN RETURNING TO THE BODY

It may appear that this theory of scale has reinstated the distinction that so much recent work has attempted to dismantle: the mind/body distinction. In this scalar configuration we become confused about what "the body" is in contrast to a conceptual apparatus—like scale—for interpreting and dissecting this body. However much we would like to return *to* the body, the body itself has already become something other than what "I" usually think. Nonetheless, arguments about the mind/body distinction often seem to suggest that the return to this body is a relocating of oneself *in* an entity.[17]

In reviewing recent material feminisms, Vicki Kirby describes this move as an attempt to expunge the remnants of an anthropocentrism that would place the "human species being as sole author and reader of its world."[18] Yet, in this reworking of the human, we find the same problem of identity: "If being is a diffracted, non-local manifestation *of* the world's Being itself, it is nevertheless, and at the same time, a local and very specific punctum, or concrescence, *of* that same world: although the individual is not in position to the universal, . . . individuation is always and necessarily unique" (18). This is the perplexing scalar shift we are examining here. However, in framing this return as bodily or material it becomes difficult to rethink the distinctions of mind/body, interpretation/matter, culture/nature that are at stake. Thus, Kirby declares the riddle to be "how should we . . . comprehend the goop and spill of corporeal interiority, the bone, muscle, and sinewy connections . . . the greens, reds, yellows, and browns that pulse and ooze just under our skin?" (19). Curiously, this gloss is noticeably nonscalar: it doesn't start with the more bewildering fact that this goop is actually a dynamic and distributed network of chemical interactions and a lively set of bacteria in your gut.[19] This return to nonscalar language permits Kirby to retain here the possessive "our" in relation to the body (they are our bodies), which in turn permits her to posit something like the very authorship that ought to be expunged as an anthropocentrism: "We need to consider if a discursive analysis *of* the body is authored *by* that same goop and spill" (20).[20] But, we might ask, by the microbiome or the ecology?

Of course, Kirby is complicating the body even as she invokes it. I am pushing at this articulation in order to identify three ways that this turn to the body can interfere even with this task of taking this flesh and bone seriously. In rescaling to these objects called "body" or "matter," we risk forgetting the essential abstracting required for this scalar shift. In response we can inquire about the scalar nature of the mind/body, information/material, and nature/culture distinctions. In each case, what is the immaterial and the disembodied?

First, the simplest point: the nature/culture distinction is a scalar distinction. In this regard, scale is quite useful for the question organizing Kirby's edited collection—"What if culture was nature all along?"—or in Evelyn Fox Keller's mirage of nature and culture.[21] When one argues against material, biological, or atomistic explanations, using arguments from the social or the cultural, one is attempting to change the scale of relevance or importance. But to suggest that bodies or subjects are socially determined is to operate under the same scalism being resisted. In turning back to the material or biological, we need not leave behind the social/cultural but understand that we are mapping different scales of influence.[22] We need to understand the scalar relation between such things as bacteria, molecules, social systems, and ecological constraints without reifying any of these as inherently determinant of the others (6.34). Instead, scale helps us understand some of the problematic maneuvers performed in science, technology, and medicine, as Annmarie Mol does when she notes the ways in which medicine makes the body multiple.[23]

Second, scale disrupts and reworks what we mean by "embodied" and "material" in a way that confuses many invocations of these terms. For instance, take N. Katherine Hayles's critique that information has been disembodied:

> Information, like humanity, cannot exist apart from the embodiment that brings it into being as a material entity in the world, and embodiment is always instantiated, local, and specific. Embodiment can be destroyed, but it cannot be replicated. Once the specific form constituting it is gone, no amount of massaging the data will bring it back.[24]

What does Hayles mean by "embodiment" other than the lower-scale substrate in which information is instantiated—the transistors or cells that carry the pattern? While cybernetics and information theory deal with "patterns rather than physical entities," this does not mean that information inherently loses its body.[25] While Hayles is right in resisting the maneuver that privileges larger-scale patterning, the return to embodiment confuses what is at stake: a difference between the patterning of components for the sake of larger-scale behavior and the *instantiation* of that pattern in a lower-scale substrate (see 6.30–34). In this regard, "embodiment" cannot be replicated, because any given instantiation of a pattern will not always translate to a different substrate. At the same time, Hayles makes clear that her return to embodiment is an attempt to resist carrying forward a humanist notion of the self: "What is lethal is not the posthuman as such but the grafting of the posthuman onto a liberal humanist view of the self."[26] Her prime example—Hans Moravec's fantasy of brain uploading—already presumes a coherent identity able to be read, copied, and transferred into another medium without fundamentally changing its structure and form. However, Hayles's return to the language of embodiment obscures how embodiment provides one primary basis for this sense of "I." While fantasies of brain uploading may glorify information apart from the body, they preserve a sense of self partially derived from the sense of being embodied.

The problem of embodied information might be better conceived as a use of scalar structures for the sake of abstraction. In this case, scale will help us, third, to specify the nature of the abstraction. We can begin to do so by again starting with a critique of this abstraction, as exemplified by an early moment in the feminist science studies conversation: Evelyn Fox Keller and Christine R. Grontowski's "The Mind's Eye." In connecting feminist critiques of disembodiment with objectivity in science, the authors posit two modern scientific assumptions about the knowability of nature: "the separation of subject from object" and "the move away from the conditions of perception." These are said to be a "retreat from the body sought in Plato's epistemology."[27]

However, this "retreat from the body" was an attempt to instantiate a systematic accounting of our mental apparatus whereby we experience reality. In Plato's *Protagoras* we see scale enter as a need to account for perceptions:

> "Do things of the same size appear to you larger when seen near at hand, and smaller when seen from a distance or not?" . . . "If then our well-being depended upon . . . doing and choosing large things, avoiding and not doing the small ones, what would we see as our salvation in life? Would it be the art of measurement or the power of appearance? While the power of appearance often makes us wander all over the place in confusion, often changing our minds about the same things and regretting our actions and choices with respect to things large and small, the art of measurement in contrast, would make the appearances lose their power by showing us the truth." (356c-e)

We need to move from perception to taking measure of our perceptions; otherwise we find ourselves changing our minds (or words) about the same things, speaking of them sometimes as atoms and molecules, sometimes as cells, sometimes as bodies, sometimes as larger ecologies. Rather than a retreat from the body, this "art of measurement" is an attempt to rebuff the presumption that the world produced by the *Homo sapiens* body is singular, coherent, and fully present.

As discussed in the last chapter, Plato's push against privileging the body is not about knowledge (*epistēmē*) but rather an attempt to open ourselves up to the various ways reality presents itself. Thus the *Theaetetus,* which Keller and Grontowski reference for Plato's discrediting of knowledge derived from the senses, is a text about the problem of considering knowledge as certain and absolute. Keller and Grontowski cite only one moment in the conversation attempting to define "knowledge," in which the interlocutors discard knowledge gained from the senses.[28] The remainder of the *Theaetetus* continues to inquire into knowledge, focusing on this incapacity to delimit oneself clearly and wholly as *the knower*. The dialogue goes on to posit *but then discard* two additional notions of knowledge that seem to be at the core of the modern scientific conception of knowledge: opinion or judgment (*doxa*) and opinion accompanied by an account or reason (*logos*). Importantly, it dismisses both of these by asking *who has* the knowledge, which undermines each definition since it is unclear who is present to know through opinion or reason (e.g., at 197b). In the end, the dialogue does not create a satisfactory definition of knowledge but produces a different result: "You will be more modest and not think you know what you don't know" (210c).[29] This turn to humility is essential. Without it, an argument like Plato's can lead to the problematic notion of objectivity that Keller and Grontowski are critiquing. The problem comes not from this mind/body distinction itself but from the inability to conceive of knowledge as not fully accessible to a coherent human entity delimited as "I." Thus, Keller and Grontowski argue that "insofar as it does not seem possible to conceive of knowledge as a passive recording of data—*human pride alone would seem to preclude such an epistemological posture*—then a sharper division between visual and mental sight was necessitated."[30] Human pride—that is, the attempt to claim the human ("me") as the knower-as-coherent-entity who has knowledge—prevents us from considering the more radical cross-scalar configuration in which something like "knowledge" emerges.

What, then, *is* the separation of mind and body that we found here in scale? Part of the difficulty is that, in Plato, the word for "mind" or "soul"—*psukhē*—is filling in for too much, including at least thought, memory, ego, and reason.[31] What we need for scale, however, is the level of abstraction above perception: a perception of perception, a capacity to measure, compare, and dissect these various perspectives and components. While this capacity may require thought, memory, and rational capacity, it does not require a sense of identification that takes this "mind's eye" and ascribes it to a *cogito*. Usefully, in Sanskrit terminology, ego is separated from thought, memory, and reason as *ahaṃkāra*, which can be translated as "the 'I' who is the doer." *Ahaṃkāra* is almost always said to be an illusion, and thus could be glossed as "the I who supposes that it is the doer." To the contrary, *cit* is attention (the capacity to attend to), and *manas* indicates more complex mental formations such as those that reflect back on and organize sense data.[32] Neither *cit* nor *manas* requires an ego, *ahaṃkāra*, the sense of you as the doer. To the contrary, Plato's *Theaetetus* suggests that egoic identification corrupts the process of attending to any concept or perception as one fails to take into account the ambiguity of the one who is taking account.

The perception of perception does not trump or eradicate the body but operates at a meta-level above it in a way that may require a scalar shift.[33] It is a way of reflecting back on, organizing, and dissecting those structures of presentation even for any apparatus we call "body." It can be confusing to call this process "embodied," since it reflects *on* this capacity to be and form bodies and therefore is not confined to or completely limited by any given object—such as the object called "brain." Without an egoic designation of subject, any perception can be said to include not just brain but eye, light, object, memory, social meaning, hormones, evolutionary vectors, and so on. In short, every perception becomes a point at which the whole Cosmos reads itself at a particular moment. This configuration is not disembodied so much as all-bodied insofar as no delimitation can cut the subject off in advance from the manifold relations that contribute to its form.

SUBORDINATING TO THE SOCIAL MARKER OF POWER

Before we go further in this thought, however, we should pause to note that many critiques of experiences such as Edgar Mitchell's are aimed at structures of control, power, or agency. The rhetorical weight of such arguments is tremendous, particularly in their turn to the political, such as in Donna Haraway's "Vision is always a question of the power to see—and perhaps of the violence implicit in our visualizing practices."[34] If we hold our "selves" intact, we can become a "subject for" or "viewer of"—a viewer that then engages in an ego-driven structure of exploitation and pleasure.

This critique makes it easy to target resemblances of the contemplative trope of a zeroing out of our identity (see 2.9). For example, take Emerson's famous statement "Standing on the bare ground—my head bathed by the blithe air, and uplifted into infinite space—all mean egotism vanishes. I become transparent eye-ball; I am nothing; I see all; the currents of the Universal Being circulate through me; I am the part or particle of God."[35] We can understand this sentiment if we read it in scalar terms. Much like Mitchell, Emerson feels himself pervading all of reality as part of a much larger entity. The visionary experience hinges on this becoming "transparent eye-ball" as a feeling of "mean egotism" vanishing. Only then is this vision made possible, not as a limited perspective called "mine" but as part or particle of the divine—the Cosmic Whole—seeing itself.

To read this vision in terms of the power of visualizing practices is to reinsert the egoic relation that is being undermined by insisting that this perspective is a coherent site of power. In cultural criticism this maneuver often arises from applications of Susan Sontag's critiques of photography and Laura Mulvey's critiques of film. Mulvey critiques a specifically this-scale male gaze directed at women in a way that corresponds to sexual pleasure.[36] Sontag applies a similar notion of the voyeur to photography as not only a form of erotic pleasure but as an indulgence in curiosity, detachment, ubiquity, and mastery.[37] Both arguments have been extended to all kinds of film, photography, or even to conceptions of nature more generally. These arguments might be summed up by one example that, although lacking in nuance, accumulates the language of such critiques: in his critique of nature images, Bart H. Welling argues that they train the eye to be "solitary, central but remote, omniscient, all-powerful, potentially violent, pleasure-taking, commodifying, an all seeing but simultaneously invisible consuming male *subject* to its marginalized, decontextualized, powerless, speechless, unknowing, endangered, pleasure-giving, commodified, consumable female *object*."[38] Here vision is turned into a hegemonic relationship in which the person viewing nature derives pleasure from the position at the expense of the object. When we add critiques of virgin nature, such as Carolyn Merchant's critique of the female nature of Western recovery narratives, we find an indictment of the idea of Nature that is essential to Emerson's argument.[39]

These arguments are a necessary critique of a certain, nonscalar way of narrating Nature, but they are not grounds for dismissing Nature as a concept—as has become popular in ecocriticism—nor do Emerson's and Mitchell's experiences fit this critique of the male gaze.[40] Mulvey's critique is about the preservation and indulgence of the very thing we're trying to deconstruct: the male gaze derives its pleasure from the "satisfaction and reinforcement of the ego."[41] Likewise, in Sontag's voyeurism the distance to the image is a way of keeping oneself intact, which permits the relation to be an indulgence and curiosity. Rather than distance and egoic indulgence, Mitchell and Emerson experience a ubiquity conditioned on their dissolution in the view. A voyeuristic approach to such views—or Nature more generally—*hinges on not implicating oneself into the view*. To the contrary, taking "this perspective" as a zero point (2.9) results in quite a different subject position: an emptying out of the power one egotistically claims within this limited human body.

SITUATED IN/OUT OF POSITION

This yielding of the remnant identification is essential for any accounting of the production of an experience. The elephant in the room here is Haraway's "Situated Knowledges" and her much cited notion of the "god trick." Like the critiques already cited, Haraway was instrumental in reworking a particularly distorted and hegemonic claim about knowledge. Opposing the domineering view of humanity and science, she critiqued this sense of transcendence and argued for a return to position, situatedness, and the body. The term "god trick" captures the general spirit: that the claim for transcendence is a trick, an illusion in which man pretends to be in a godly position:

> I am arguing for a politics and epistemologies of location, positioning, and situating, where partiality and not universality is the condition of being heard to make rational knowledge claims. These are claims on people's lives. I am arguing for the view from a body, always a complex, contradictory, structuring, and structured body, versus the view from above, from nowhere, from simplicity. Only the god trick is forbidden.[42]

We can see how this intuitive reference point of "my" experience would provide an underlying sensibleness and weight to Haraway's distinction. Any perspective said to be beyond this situated, bodily perspective would seem to generalize out this experiential configuration. Such a generalization would not only be problematic; it would be skewed, partial, and not very useful for science. While I am not arguing here against situating knowledge claims, we see in scale that there are crucial ambiguities about defining and delimiting this situatedness. In scalar views there is something like a transcendent view that moves away from the body itself (2.10–12), zeroes out the perspective as "mine" (2.9), and dislocates from any single perspectival configuration (2.15). Thus, when Haraway declares that "feminists don't need a doctrine of objectivity that promises transcendence" (579), we ought to hesitate. Does scale not provide us with something like a transcendence that is required to account for various modes of experience?

I'd suggest that Haraway is targeting a particular kind of transcendence that should be distinguished from an earlier, contemplative "view from above/nowhere." This contemplative transcendence is described, for example, in the anonymous fourteenth-century text *The Cloud of Unknowing,* as a becoming-nowhere, but not as a place of impeccable and impenetrable self-hood:

> Do *not* try to withdraw into yourself, for . . . I do not want you to be anywhere; no, not outside, above, behind, or beside yourself. But to this you say: "Where then shall I be? By your reckoning I am to be nowhere!" Exactly. In fact you have expressed it rather well, for I would indeed have you be nowhere. Why? Because nowhere, physically, is everywhere spatially.[43]

This view is already necessary for scale; if we want to be able to work with and acknowledge the multidimensionality that is already implied by scale, then we have to dislocate ourselves from the perspective already said to be situated "here" in the body, at this location, in this single form (2.8).

However, this account is not, I'd argue, Haraway's primary target. Instead, the god trick is more properly a *failing* to zero out in this way. The target text is more appropriately Thomas Nagel's *The View from Nowhere,* published shortly before Haraway's essay. Nagel's argument hinges on the capacity to set aside a located perspective in favor of a more objective "view from nowhere." Here is the primary passage of interest:

> The basic step which brings [the view from nowhere/the objective self] to life is . . . simply the step of conceiving the world as a place that includes the person I am within it, as just another of its contents—conceiving myself from outside. . . . So I can step away from this unconsidered perspective of the particular person I thought I was. Next comes the step of conceiving from outside all the points of view and experiences of that person and others of his species. . . . That is the beginning of science. And again it is I who has done this stepping back, not only from an individual viewpoint but from a specific type of viewpoint.[44]

One can see the problem with this seemingly scalar articulation. This "step away" is an artificial claiming of a perspective that speaks for or contains the perspective of others. This considering "others of his species" becomes the speaking for these "others" who are not allowed to speak themselves. The possibility of "conceiving myself from outside" is then a trick granted by the privilege of a decidedly limited perspective of a subject who, not already being set aside as an other, is able to claim the process of knowledge.

Yet Nagel's view from nowhere is explicitly *not* what we just saw in *Cloud*. Thus, Nagel states that "I know this sounds like metaphysical megalomania of an unusually shameless kind. Merely being TN isn't good enough for me: I have to think of myself as the world soul in humble disguise. . . . I am not saying that I individually am the subject of the universe: just that I am *a* subject." In disavowing the nondual position, Nagel has retained his identification with the delimited position he labels "TN." He fails to see that it is *more egoic* to retain that position, since it leaves this "I" intact ("again it is I who has done this stepping back"). The result is that the maneuver is strictly imaginary—he proceeds "*as if* from nowhere"—and therefore generalizes a delimited and partial perspective while nonetheless identifying himself with this delimited experience.[45]

When Haraway calls out the trick, she attempts to undo the contradiction by reaffirming the place of the body as the source of perspective, returning the "I" back to the eye. But in doing so, she carries forward the same identification: assuming that "I" am located in a clear entity situated in a single, definite, and located way. Thus, she invokes the body first: "We need to learn in our bodies, endowed with primate color and stereoscopic vision, how to attach the objective to our theoretical and political scanners in order to name *where we are and are not*" (582, emphasis added). The task of naming "where we are" still runs into the problem of who "we" are to begin with. Recognizing this risk, Haraway moves away from the body as coherent even as she invokes it, stating that "feminist embodiment . . . is not about fixed location in a reified body . . . but about nodes in fields, inflections in orientations, and responsibility for difference in material-semiotic fields of meaning" (586). This passage deflects the question of the body's clear constitution into a sociopolitical network. In doing so, Haraway is able to disavow the need to know who exactly this subject is, while insisting that our knowledge be situated and held in reference to discernible structures of power: "We are not immediately present to ourselves. Self-knowledge requires a semiotic-material technology to link meanings and bodies. Self-identity is a bad visual system" (585). In denying self-identity but retaining an insistent situatedness, Haraway leaves only the external determination of that situatedness, a conceiving of one's position based on how it functions within a network of relations. No longer a matter of where "I" am, the "situated" location of knowledge becomes a determination within a community of others that is always partial: "The knowing self is partial in all its guises, never finished, whole, simply there, and original" (585). And yet the "knowing self," however partial, is still a self to be situated.

In some ways, Haraway's insistence that we "see from both perspectives at once" seems to fit with what scale requires.[46] And yet Haraway's damning characterization of the view dislocated from the position of the body runs counter to anything that scale produces. We thus need to be wary of this potentially inverted version of Nagel's argument. Like Nagel, Haraway risks wanting the self in both ways, at one point affirming the clearly definable and locatable nature of one's "situated" position, at another invoking what exceeds that position and thereby determines it. But then how is this knowledge situated? By others?

We pointed to scale as a kind of situated dislocation (2.18) in order to respond to this ambiguity. Scalar perspectives can hardly be called "mine" in the usual sense. The problem is that we are already so entangled with such perspectives that we do not usually account for this problem of identification as central to our assumption about what it would mean to situate "my" perspective. In turn, the body (as a this-scale object) cannot go to another scale in the sense of acting on directly on another scale (3.34). But the mind (i.e., mental apparatus) can and does—in fact it is already made up of and connected into those scalar domains of relations. However, to get to this view, we must first expunge this delimitation of this mind as inherently here and mine. Thus, self-identity is a bad visual system if by "self-identity" we mean egoic identification within any limited entity. But the complete *undoing of self-identity would look like a view from nowhere.*[47]

Only when the "I" is zeroed out in this way, even becomes the Cosmos itself, do we arrive at the zeroed-out position that includes the contingency and openness necessary to begin measuring any experience. Thus, in the centuries-long and complex debate over the nature of the "self" in Buddhism and Vedanta, Shankara stands out for arguing that the crown jewel of discrimination is the inquiry into the difference between the egoic self-identification and non-egoic identification with the Whole (Brahman). Insofar as we leave that "self" in any limited configuration, what is wrapped up in that "self" is the unaccounted-for distortion permitted to preferentially skew how we situate an experience or knowledge. *These remnants of identification produce the god trick in which a limited sense of self is smuggled into a view beyond its limits.* In responding to Buddhist refusals of any articulation of a self (*anātman* versus the *ātman*), Shankara points to identification only for the purposes of reidentifying with the Whole. This becomes necessary because the operation of reconciling "this" experience with the Whole is an inherently difficult scalar operation of moving from the *jīva*—the seemingly separate body—to the *ātman* as *brahman* (Whole) experienced here, through an active expunging of the ego (*ahaṃkāra*).[48]

SYSTEM IDENTIFICATION AND CLOSURE FROM ENVIRONMENT

We can now begin to see a scalar configuration of subjects. However, even a scalar view of subjectivity might still carry forward this base assumption that individuals are separate, in advance, via some notion of the "organism." In response, one more theory is worth examining: the theory of autopoiesis, which was developed by the biologist Humberto Maturana and the philosopher Francesco Varela and has made its way into broader theoretical conversations primarily through the German sociologist Niklas Luhmann. An autopoietic system "is a network of processes that produces the components that reproduce the network, and that also regulates the boundary conditions necessary for its ongoing existence as a network."[49] This theory positions individuation as a result of an ongoing maintenance of a system by that same system. In one sense, it keeps in view two different scales: the scale at which the individual is discerned (e.g., the body of a *Homo sapiens*) is connected to the scale at which change is continually occurring (the cellular/microbial). In fact, Maturana came to the theory of autopoiesis because he sought a "language that would permit me to describe an autonomous system in a manner that retained autonomy as a feature of the system."[50] Therefore, the organism is said to be individual because it maintains the system of lower-scale interactions that stabilize as a body on a different scale.

This individuality is both mechanistic and based on a notion of the network.[51] Despite the scalar nature of the description, we can ask: does the system as a system treat itself as a coherent entity set apart from the rest of reality? The sticking point is always going to lie in the extent to which the system must inevitably be "open" (the theory emphasizes the closed nature of the system) to that which is outside of it in order to maintain itself. There is always a residual porousness that permits what appears to be foreclosed on one scale to be revealed as interpenetration at another. Just change the scale one more time and individuality becomes less important than porousness, in how the ecosystem, for instance, is taken in and used and reworked by the body of the *Homo sapiens*. From which scale is this system said to be an individual? Wolfe points to this as a paradox in his discussion of autopoiesis: "the functional distinction system/environment" run up against the "means of autopoiesis and self-referential closure as a means of self-preservation."[52] However, our discussion suggests that this paradox is only created if we assert that individuality is an actual designation rather than a functional, partial, and secondary delimitation.

Crucially, Maturana and Varela entangle this problem with questions of mind by arguing that autopoiesis also creates a phenomenal domain (xxi). Thus they argue against "evolutionary and genetic notions" that "treat the species as the source of all biological order," since such explanations do not "provide us with a mechanism to account for the phenomenology of the individual" (115). In fact, they argue that there is the need to provide "a mechanism for the definition of the individual" because we need to define and account for these phenomenal domains. Maturana and Varela thus insist that individuality is necessary for ascribing a phenomenological domain *to* an organism, which is treated as a pre-given unity; they simply assert "that machines are unities is apparent" (77). In turn, a "phenomenological domain is defined by the properties of the unity or unities that constitute it, either singly or collectively through their transformations or interactions" (116).

Do we have to assume the unity of an organism to account for the specificity of a phenomenal experience? I'd like to propose that this argument is backwards. It still assumes, in advance, a clear individuated organism producing experience. In contrast, we need to start from a simpler point: an experience occurs. This experience might be considered constitutive of what we call a "mind," but it is not clearly and inevitably tied to a particular system. Only with an additional observation do we call this system a unity as we attempt to explain the configuration of reality that produces that experience. In contrast, Maturana and Varela assume the unity not only of the autopoietic system but also of the observer: "Anything said is said by an observer" (8). But how do we move from the particularity of experience to the assertion that the observer is a separate, coherent, individual reference point? Only in an additional observation that says either "this phenomenal realm is me" or "that (organism) is an observer."

To get to this point, one has to ask one more time, as does fellow second-order cyberneticist Heinz von Foerster, "Who sees this observer?"[53] Foerster turns the question not to the external "observer" but to the observer here observing the observer. On this basis, Foerster notes that, while he always found autopoiesis a "very significant definition" of life, this definition operates with too much assumed. Specifically, it is "centrally concerned with what would then be a living system."[54] In other words, the organism is presumed first and then autopoiesis and phenomenal domains arise from that assumption. But, as Gilbert, Sapp, and Tauber argued, assuming individuality is not as useful even for biology as we might suppose. Foerster's position on identity is more recursive: "For me, 'I' is a folded up recursive operator of infinite depths, but one can operate with it."[55] This may be more appropriately functional, but only if it leaves open the possibility that the sense of self is *an additional delimitation within an experience that has already been generated.* Only in the recursive observation of the observer observing itself does the system get defined as a system, an "I", an individual (more on this in chapter 12). The question will always be: do we, in practice, really leave this defining process open?

The openness is particularly significant because of the potential relationship between scale domains and phenomenal domains. If we fail to acknowledge this secondary nature of the designation of "self," we find ourselves with a significant omission that prevents us from, among other things, understanding scale. After all, scale requires that we admit that within this phenomenal experience is already the action of things not discernible, but which nonetheless contribute to this system both in terms of action and experience. We risk leaving out in advance additional content and qualia already in an experience but not identified (either by this self-referential thought or by others) as being part of the system. Here we find the strange twist that is expressed, for example, in Andy Clark and Chalmers's extended mind thesis: because these phenomenal domains are identified with the body, it is surprising to argue that cognition and other forms of mental awareness inevitably happen outside these boundaries.[56] In turn, the long history of speaking in terms of the unconscious does the same for other scales. But within your phenomenal domain is already the "unconscious" of the cellular, the ecological, and so on. We become confused at how these are also a part of "me" because we do not recognize that "I" am already those things, even if they are not direct contents recognized in experience. What if we must scrub ourselves of this primary self-identification in order to see within our "selves" this scalar situation?

PROBLEM OF SELF, PROBLEM OF EXPERIENCE

Even though scale pulls our experience out of the scale of the body, we still have to explain its relation to the here of experience. These examples have shown that we need to separate out the problem of mind/experience from the problem of identification/self. Theories of mind, particularly Chalmers's hard problem of consciousness, point to how structures of mind produce an experience not reduced to those components. Yet this problem of consciousness is also a problem of positionality or situatedness—an accounting for and specifying how an observation is created by and out of a particular nexus of factors for parsing reality. The problem of identity haunts this whole question, because identification is not the same as situatedness. Identification requires an additional delimitation of that experience. To put it in Nagel's phraseology: we might acknowledge that there is something it is like to have a particular experience and then, apart from that, is the description of who or what produces that observation (e.g., a bat). The whole confusing conundrum lies in accepting the lack of definitiveness of the observer or apparatus even as a particular observation is presented with certain attributes, providing certain information, and presenting grounds for knowledge (*epistēmē*) and response.

To summarize the thesis most axiomatically: *epistemological specificity does not equal ontological separability.* Just because we find particular experiences that are produced from a unique configuration of reality, and just because these require careful accounting to say anything about how they are produced and what they apply to, it does not mean that these experiences are attributable to coherent and contained subjects. The problem repeated in the previous sections is that this axiom is inverted or avoided: either we assume ontological separation first and use it to account for epistemological specificity, or we take epistemological specificity as evidence of ontological separation. Much remains to be said about the grounds for tracing epistemological specificity such that we can say anything meaningful at all about the production of experiences without assuming individuality in advance. We do not need to propose such a theory of mind here. I only want to posit that *first starting with criteria for mind permits us to designate phenomenal boundaries before we provide physical ones.* Such a formulation would be more consistent with this scalar configuration toward subjects and provide us with some grounds for acknowledging the specificity and uniqueness of experience without claiming it for and by a particular segment of reality to be called "subject."

FINDING TRANSHUMAN CONSCIOUSNESS

What, then, do we make of this "subject" Edgar Mitchell, floating up in space and becoming larger than himself? In a real sense, he has scaled beyond himself, feeling and seeing that what is already within his experience is far more than might be attributed to this little body. In this situation, we might scale back to here and reconsider anew the human gaping out into the scalar vastness. If we do not merely reassert that locality but instead attend to how that locality is reconfigured in scale, then we find ourselves with quite a different perspective on this whole prospect of perspective. The human is not left behind but newly placed and positioned in this extended and layered reality of which it has now become aware.

Properly speaking, such an account would not be a posthumanism (even if it pushes past what is called "humanism") but a transhumanism. As Andrew Pilsch has made clear, the term *transhumanism* is not merely the recent techno-fetishization that yields notions like brain-uploading of the sort that Hayles critiques. As Hayles demonstrated in her reading of cybernetics, such philosophies are plagued by a remnant fantasy of the humanist subject even as that subject is deconstructed by the very technologies supposed to preserve it. In Pilsch's account we see an alternative and perennial tradition that transforms this human subject: a "mystic transhumanism" or "cybernetic mysticism."[57] When Pilsch introduces the Jesuit anthropologist Teilhard de Chardin as central to this mystic transhumanism, it is not brain uploading or fantasies of living forever that are emphasized but a radical "reorientation of consciousness" (125) that hinges on the "ability of consciousness to reflect on itself" and continue to expand this conscious reflection (126). A transhumanism that denies these mystical elements—for example, in Nick Bostrom's desire for a transhumanism "without resorting to wishful thinking or mysticism" (121)—risks preserving this self as the ego-who-transcends.

In invoking transhumanism, I want to further Pilsch's suggestion that the way to understand our transhuman condition "is not [through] more data or more gadgets but a more mystic transhumanism" (61). While some posthumanisms turn away from the human, a mystic transhumanism reorients the human subject within the larger possibilities of experience that we encounter in scalar descriptions. Thus, Franklin Merrell-Wolff, in discussing the possibility of transcending Kant's a priori limitations in consciousness-without-an-object (see chapter 7 here), stumbles upon transhumanism as a label for this configuration:

> It may be valid enough to assert that human consciousness qua human is always [conditioned, e.g., by time], but that would amount merely to a partial definition of what is meant by human consciousness. In that case, the consciousness that is not time conditioned would be something that is transhuman or nonhuman. . . . It is in the power of man to transcend the limits of human consciousness and thus come to a more or less complete understanding of the factors that limit the range of human consciousness qua human. The term "human" would thus define a certain range in the scale of consciousness—something like an octave in the scale of electromagnetic waves. In that case, the present system implies that it is, in principle, possible for a conscious being to shift his field of consciousness up and down the scale. When such an entity is focused within the human octave, it might be agreed to call him human, but something other than human when focused in other octaves. . . . On the basis of such a definition, this philosophy would not be a contribution to Humanism but to Transhumanism.[58]

In this articulation, the human is defined by the ability to become aware of its own limits, to trace extensions beyond these limits, and to situate its present experience within these dislocations. Insofar as "human" indicates a limit and configuration of experience—the *Umwelt* or lifeworld, as biologist Jakob von Uexküll once labeled it—of the *Homo sapiens*, then the human is one register of the possible ways of encountering, experiencing, and dividing the world. Perhaps, we can call this "human register" reality viewed at the scale of the meter and filtered through the particular differences relevant to this organism. However, even in this definition, scale makes us aware of that which exceeds this delimitation. Thus, Indian philosopher Sri Aurobindo describes the individual in these terms: "As the movement is one and indivisible, so he who is present in the movement is one, sole and indivisible. . . . What then must be the spiritual position of the personal worker? . . . He is a center only—a center of differentiation of the one personal consciousness, a center of determination of the one total movement."[59] Here the posthumanist decentering of the human is turned inside-out as any structure of mind is made a center—not as *the* center but rather a figure and shadow out of which the Whole experiences itself (see 5.18).[60] *I* cannot be this figure or center, insofar as anything can be kept here as *ahaṃkāra,* the one who pretends to be apart from the One-Whole-All.

In fact, this ego-structure will never fully function because what it attempts to claim is already a shadow reflected within this All. Everything short of arriving at the Absolute is a cut made without truly accounting for this cut performed as "oneself." As the twentieth-century sage Ramana Maharshi argued, the first cut is always between "I" and the rest of the cosmos. That is, the ego-structure is the first partial-whole (1.28) designation (the subject) out of which all other partial-wholes (objects) are identified (see 2.24). Examining this sense of the "I" is the necessary condition for getting beyond subjects and objects: "The ego's phenomenal existence is transcended when you dive into the source from where the 'I'-thought [*aham-vritti*] rises."[61] However, this transcendence does not eradicate all sense of selfhood, but only the sense of self as anything less than the Whole: "You must distinguish between the 'I', pure in itself, and the 'I'-thought. . . . If you stay as the 'I', your being alone, without thought, the 'I'-thought will disappear and the delusion will vanish forever."[62]

At the same time, if we follow this logic carefully, the transcendence is not a transcendence at all but a returning to what one truly is: the Cosmos looking back at itself. Otherwise we are forced to ask, with Maharshi, "Transcending what, and by whom? You alone exist."[63] But we can augment this "you alone" with the careful negation, drawn from the Ribhu Gita, one of Maharshi's favorite texts: "I am not the individual. I have no differences. I am not thought nor do I have any mind. I am not flesh nor do I have bones nor am I the body with ego [*ahaṃkāra*]."[64] Edgar Mitchell feels his flesh and bones extending out into the cosmos. He suddenly feels certain "scientific facts about stellar evolution [take] on new significance" as he feels how this body was churned out from the unknown abyss.[65] A billion suns explode over billions of years to evolve out bits of starstuff that rocket themselves up to space to get a better view of the vastness. This bundle of cells feels, looking out, torn apart, echoing the Ribhu Gita: "ever of the undivided nature am I; ever devoid of measure. Ever of the pure and singular form am I; ever the consciousness alone" (22.29). The measure by which this world is measured here is itself not the Absolute it persistently measures. Yet the Absolute measures itself, in this form here to be illuminated within itself: "Ever of the nature of true measure am I; ever the illuminator of existence" (22.30). And yet, and yet, we pause once more for this limit to be exceeded: "I am not the one who measures. I am not that which is measured. I am not all. I am the supreme. I am the form of complete knowledge [*sarva-vijñāna-rūpo*]" (22.35).

Thus, in examining this "I-thought" we find the transcendence from part to whole and back again to this here and now. This transition has the power to eradicate the sense of absoluteness granted to the temporary and the partial, the permanence we grant to these shifting forms, and the hold by which this realm of appearance binds us. Without this reconfiguration of identity, the view held as "mine" is always dark and partial. In this way, we stumble into Paul's first epistle to the Corinthians: "For we know in part, and we prophesy in part. But when the perfect [*teleion*, the complete in all its parts] comes, the partial will come to an end." Everything done by this body and mind is partial, but in examining itself, this partiality already sees into its own partiality. "Now I know in part, but then I will know fully, as I am fully known."[66]

This strange Möbius strip is depicted in the film *Interstellar* when, in search of a solution to save the human race, the interstellar astronaut named Cooper falls into a black hole. There he finds a replica of his daughter Murph's bedroom repeated endlessly around him, each iteration showing a different time of the same place. He realizes that this is some kind of interface by which he can transmit data about the black hole gathered by Tars, the robot who also fell into the black hole, through space-time back to his daughter on Earth so that she can complete their antigravity spaceship. As he realizes what he needs to do, he suddenly understands why he must be there to do it. Speaking of the mysterious "they" who made this trip possible, Cooper and Tars finally realize their situation:

> COOPER: All of this is one little girl's bedroom, every moment, infinitely complex. They have access to infinite space, but they are not bound by

> anything. If they can't find a specific place in time they can't communicate. That's why I'm here, to find a way to tell Murph, just like I found this moment.
>
> TARS: How, Cooper?
>
> COOPER: Love, Tars. Love. . . . My connection with Murph . . . it is quantifiable. It's the key.
>
> TARS: What are we here to do?
>
> COOPER: Find out how to tell her.

Precisely because "they" are beyond space-time—in some way infinite or exceedingly vast—they need a limited entity to go into the black hole and take the measurements. Cooper and Tars are the devices by which that precise moment is found and communication is made possible. This human circuitry is always just this: a means for the Cosmos to see, feel, think about, and do just what is right here—as a navigation of itself particular to the limits of this human configuration. Once we realize this—"that's why I'm here"—we are open to do the task at hand, to find a way to tell Murph, to find a way to do whatever it is we need to do. (But how? Here we struggle to find the word to articulate the sense of already being drawn and connected into this reality—why not call it love?) After this recognition, the ingenuity by which this task is carried out naturally emerges from the mental and perceptive capacities already present in the delimitations of our experience. However, this whole perspective can only be acquired if we ourselves go through the black hole, into the abyss in which our sense of egoic self is lost. Only in this involution do these fundamentally transhuman views manifest as a reorientation wherein the limitations of this flesh and blood are suddenly imbued with a renewed purpose and meaning. Then the mouth opens to speak, the hands find themselves writing, the breath continues to intake the cosmos, the careful judgment measures the perception at hand, and we move forward through space-time.

9

Cutting and Claiming Everything

Scalar Configurations of Relations

THE EARTH FROM THE HEAVENS

Rusty Schweickart finds himself circling the planet. In this circumambulation, he does what most humans seem to do: he finds what he recognizes. First, upon waking up, North Africa, and then, after breakfast, the Mediterranean, and so on. In this circling, his initial insight is about this civilization: "And you realize in one glance that what you're seeing is what was the whole history of man for years—the cradle of civilization."[1] A fine enough conclusion, but one still predicated on familiarity and human relations. Thus, as he moves along, he finds "finally" the "coast of California and look[s] for those friendly things"—"there's home."

But he doesn't just do this once, but "again and again and again." After a while he finds himself experiencing a gestalt shift: "When you go around it in an hour and a half you begin to recognize that your identity is the whole thing. And it makes a change." This change culminates in what he famously describes as "there are no frames, there are no boundaries." These forms of human conquest, powers, and priority are revealed as spurious and temporary impositions on this large-scale entity. Suddenly, this scalar view provides a new sense of priority and a new sense of relations: "That relationship [with the planet] is no longer what it was." Schweickart doesn't stop there but, invoking his peers who left near-Earth orbit, adds:

> And now he looks back and he sees the Earth not as something big, where he can see the beautiful details, but he sees the Earth as a small thing out there. . . . The size of it, the significance of it—it becomes both things, it becomes so small and so fragile, and such a precious little spot in that universe, that you can block it out with your thumb, and you realize

that on that small spot, that little blue and white thing is everything that means anything to you.

This now familiar trope has been repeated in numerous contexts, from environmentalism slogans to existential musings in popular television. This trope uses a higher-scale vision turned back on the Earth to input into that vision the whole field of relations and priorities on which human life is built. The result is a diminishing of their significance for the sake of reworking their importance within the larger Cosmic context. But this vision is also balanced by this return to the specificity: it becomes, not just insignificant but a "precious little spot."

Of course, this trope has a deep history prior to space travel. We can, as an introductory comparison, sample two additional examples. In her "showings," Julian of Norwich experienced a noticeably scalar vision: "And in this vision he showed me a little thing, the size of a hazel-nut, lying in the palm of my hand, and to my mind's eye it was as round as any ball. I looked at it and thought, 'What can this be?' And the answer came to me, 'It is all that is made.'"[2] The change created by this vision is written now as a result of a limited human perspective: "This is why those who choose to occupy themselves with earthly business and are always pursuing worldly success have nothing here of God in their hearts and souls; because they love and seek their rest in this little thing."[3] We could likewise refer to the essential passage from the Bhagavad Gita, in which Arjuna encounters the true form of Krishna: "There Arjuna then beheld the entire universe as One, divided in many ways in the God of Gods. Filled with amazement, hair standing on one end in ecstasy, he spoke these words: O Lord, I see within your body all kinds of beings assembled."[4] The basic maneuver is the same: a larger-scale reference point is invoked that not only diminishes but includes and reorients all of the objects and relations available within human experience. The result for Arjuna, as for Schweikart and Julian, is a whole new relationship to the reality around him as he is called forth to the familial battle framing the *Gita*.

I put Schweikart's speech next to Julian and the Bhagavad Gita to highlight this emphasis on transformation of values and relations. But it also highlights the distinctively contemplative nature of the experience. Here we run into another persistent trope of the Space Race seemingly in tension with this contemplative experience: that this vision is a leaving behind and a forsaking of the important concerns of this planet.[5]

For example, in the opening reflections of *The Human Condition,* Hannah Arendt responds to the excitement over the launch of Sputnik with this critique:

> The immediate reaction . . . was relief about the "first step toward escape from men's imprisonment to the earth." And this strange statement, far from being the accidental slip of some American reporter, unwittingly echoed the extraordinary line which, more than twenty years ago, had been carved on the funeral obelisk of Russia's great scientists: "Mankind will not remain bound on this earth forever."[6]

There is undoubtedly something in space travel that "seems to be possessed by a rebellion against human existence as it has been given" (2). Arendt's response might always arise in encountering Schweikart, Julian, or the Bhagavad Gita: but what about the Earth?

Arendt's critique focuses on the scale of *Homo sapiens,* focusing primarily on alienation from this-scale existence. Thus she places the Space Race in the context of three other major world events: the colonizing of the Americas, the Protestant Reformation, and the invention of the telescope, which made possible a "science that considers the nature of the earth from the viewpoint of the universe" (248). The result of this viewpoint, she argues, is alienation:

> The greater the distance between himself and his surroundings, world or earth, the more he will be able to survey and to measure and the less will worldly, earth-bound space be left to him. The fact that the decisive shrinkage of the earth was the consequence of the invention of the airplane, that is, of leaving the surface of the earth altogether, is like a symbol for the general phenomenon that any decrease of terrestrial distance can be won only at the price of putting a decisive distance between man and earth, of alienating man from his immediate earthly surroundings. (251)

Arendt rightfully identifies a basic problem with scale: "But it could be that we, who are earth-bound creatures and have begun to act as though we were dwellers of the universe, will forever be unable to understand, that is, to think and speak about things which nevertheless we are able to do" (3). We are now able to do things that extend beyond our terrestrial lives and the limits of these human bodies: this is a fundamental implication of scalar science and technologies. However, Schweikart's return to the preciousness of the Earth—indeed the entirety of the scalar discourses we are examining here—makes one wonder if, in fact, we will forever be unable to think and speak about these things. As we have noted (2.8), scale always is a relation between two experiences, one of which will include the here. The implication is that we would only be alienated from this-scale experience or be unable to speak of it if we do not properly understand these views and actions in terms of scale.

This clash of tropes—between the discovery of insignificance and the reassertion of values derived from this-scale existence—points to a need to reexamine how scale morphs human views and values by pushing them beyond the human. The problem of alienation is essential for highlighting how one can forget, in a scalar view, that *Homo sapiens* bodies dwell on the surface of the Earth. But the greater point is that we must rework our sense of the human condition around this new scalar perspective. Arendt, in her attempt to return us to the Earth, opens up the question but, without a conception of scale, is unable to see the new map of relations that scale makes possible.

This chapter will examine this tension between classical or intuitive ways of mapping relations and those encountered in scale. Here, I want to focus on another trope—synecdoche, the part filling in for the whole—to examine how scalar remappings of relations come into tension with nonscalar vestiges of thought and articulation. This chapter will introduce this trope in relation to the scalar views from above, then consider how synecdoche highlights the primary scalar tension in our way of conceptualizing relations. We will then examine this tension in the humanities and social sciences, specifically geography, sociology, and political theory, by examining some critiques of this synecdoche that have been, curiously, situated as critiques of scale. We can then apply the same questions to the sciences, examining the scalar synecdoche as it emerges in systems theory and related sciences as well as nanotechnology. We will then return to this tension between the contemplative and the active, the here and the scalar, to consider how such remappings are also implicated in the personal alteration of one's system of values and one's relation to the Cosmos.

DREAMS OF TRANSCENDENCE: THE COLONIZING AND THE TRANSFORMATIVE

The *Homo sapiens'* experience of the world is rooted in a particular scale of perception and filtered according to its needs (whether biological or social). This perspective becomes normalized within our language and thought. But even the simple perceptive variances created by distance (1.1–3) already confound our ability to render this experience coherent. Thus, scalar views have long been used as a trope for reorienting human relations, even prior to the scientific extension and systemization of scale. Many such articulations were mystical in character, since they spoke of a reality that exceeded the human. Others were political in nature for the same reasons that we found in 3.9: the aggregation of human efforts is one of the most readily available means of considering the coming-together of smaller objects into a larger unity. Sometimes both purposes occur together, as they do in Plato's *Republic*, where Plato discusses the republic in order to discuss the ruling of the soul, as a kind of scalar isomorphism (see 369a). Similarly, the Bhagavad Gita's setting in the battlefield puts the transcendental experience in the context of the necessities of politically charged action.

These contemplative and political invocations of scale are often conflated, but I want to separate them here in order to highlight an essential difference in their form and effect. Depending on whether and how the human is expanded into, preserved, or reworked in such scalar views, one can distinguish the contemplative, transformative invocation of scale from the political, controlled vision of scale. Contemplative scalar articulations transform this-scale experience, using scalar views to pull us out of, make strange, and rework our priorities. Political scalar views tend to extend the human into realms that exceed the *Homo sapiens* body by attempting to map, gather, and claim these larger views as a structure of power. This practice is best represented by mapmaking, which scales drawings according to known attributes of a landscape and focuses on the objects we hold important: cities, landmarks, roads, rivers, and, most importantly, political boundaries. These two scalar views exist in tension and cannot be reconciled, since one attempts to claim the scalar view in this-scale terms while the other rewrites this-scale reality in a scalar view. Even those that position the contemplative view in relation to the political, such as Plato's *Republic* or the Bhagavad Gita, must be distinguished based on whether and how the human is reworked in the view.

Consider another precursor to Schweickart's vision: Cicero's *Dream of Scipio*. In this work, the Roman general dreams of ascending to the heavens where he has a conversation with the deceased Africanus. In this narrative, colonial desire is positioned as a resistance to transcendence: Scipio resists bringing his gaze away from the Earth, leading Africanus to declare: "Come! . . . how long will your mind be chained to the earth?"[7] Nonetheless, Scipio persists in focusing on the Earth, leading Africanus to declare: "For all the earth, which you inhabit . . . is a kind of little island surrounded by the waters of that sea . . . and yet though it has such a grand name, see how small it really is! . . . you can certainly see in what a narrow field your human glory aspires to range."[8] Here, in the midst of Roman imperialism, is a declaration of the futility of empire, of the smallness of any possible actions that seek to dominate the Earth. This view is thus antithetical to the colonizing, mapping enterprise; it requires a turning away from the details of the surface that one would like to dominate for the sake of reworking this sense of value and priority that is the impetus for that very domination.

However, it is easy to dismiss the more transformative, contemplative encounter on the account of the political. For example, in his archaeology of representations of the Earth, Dennis Cosgrove places the two elements together as if they are unavoidably entangled: "The history of such representations is complex, connected as closely to lust for material possession, power, and authority as to metaphysical speculation, religious aspiration, or poetic sentiment."[9] While he appears to separate out the two, Cosgrove subordinates possible transcendence to colonial desire:

> The dream of human flight sufficiently high to offer a global perspective is an enduring theme of stoic philosophy. . . . This is the implication of whole-earth literature from Cicero, Lucian, and Seneca which offers its male heroes their destiny in synoptic vision. Their *telos* combines an imperialistic urge to subdue the contingencies of the global surface with an ironic recognition of personal insignificance set against the scale of globe and cosmos.[10]

Cosgrove distinguishes the two aspects only to conflate them, ignoring that these writers were explicitly critiquing the colonial enterprise. He deems this recognition "ironic," but, except in that Lucian's *A True Story* is a satire, all three are quite serious about the sentiments instilled by their visions. This critical move imposes onto these texts the very thing being critiqued, reinscribing forms of difference only discernible at the scale of human bodies and imposing them onto the scalar experience. In response, we need to more carefully trace how and to what extent we have succeeded in following the contemplative revision of the human rather than the colonizing extension of it. This difference is a difference in effect (what does the articulation lead us to think and do?) that might be traced out in the history of political theory, technological and scientific articulations of control, and our use and misuse of contemplative transformations.

SYNECDOCHE AND THE CLAIM FROM ABOVE

Let's first examine how the human comes to fill in for other scales of relations by examining this tension in NASA historian Frank White's attempts to articulate the power of the perspective granted by space travel. In outlining what he calls the "overview effect," he first invokes the experience of the airplane:

> From 30,000 feet, [the city] looked like little toys sparkling in the sunshine. From that altitude, all of Washington looked small and insignificant. However, I knew that people down there were making life and death decisions on my behalf and taking themselves very seriously as they did so. From high in the jet stream, it seemed absurd that they could have any impact on my life. It was like ants making laws for humans.[11]

White uses this view to note how a change of scale equals a change in value: what seems significant is seen to be insignificant. Then, remembering that lower-scale set of values, White reflects on how absurd it is that people would think they have such great power over this vast landscape—they are simply too small. It follows that the power associated with these individuals is an erroneous conception: the assumption that an individual might control all of this is like assuming ants form laws for humans.

Here we encounter a nonscalar assumption: that the larger is more powerful than the smaller. This assumption is a staple of science fiction, as in the opening of H. G. Wells's *War of the Worlds*: "as narrowly as man with a microscope might scrutinize the transient creatures that swarm and multiply in a drop of water," man too is being scrutinized by the Martians.[12] Then, when the Martians attack, they "took no more notice . . . of the people running this way and that than a man would of the confusion of ants in a nest against which his foot has been kicked."[13] White's invocation flips this trope: the humans aren't controlling the ants; instead, the "humans" are the larger world viewed from the airplane. The absurdity is that a smaller organism (the ant) could control a larger one (the human). Of course, this ant-human distinction is noticeably nonscalar, since it does not cross a scalar threshold. In more properly scalar form, the analogy would be the absurdity that a single human could control the planet or that a single cell could control your body.

In these terms we discover the problem of synecdoche, taking the part for the whole. Scale naturally produces synecdoche when we attempt to situate a larger-scale view with a smaller-scale view. When in the middle of a city, I might take any part as representative for the city because this "city" is already too large and too abstract to clearly discern. However, it is only when we stay at the same scale that we can use any one of these objects to fill in for that larger-scale category, since at that new scale this synecdochic object is no longer visible. A scalar synecdoche thus inappropriately subsumes a higher-scale view to a lower-scale object. This synecdoche is pervasive within political rhetoric and conceptions of power; just as we replace the entire power structure of the United States with "Washington," the politicians there place themselves in a synecdochic relationship with this geographic region. A fully scalar view challenges this synecdochic assumption. White points to this conclusion in a definitely contemplative register: "From the airplane, the message that scientists, philosophers, spiritual teachers, and systems theorists have been trying to tell us for centuries was obvious: everything is interconnected and interrelated, each part a subsystem of a larger whole system."[14]

The view of this "whole" is not equivalent to the attempt to claim or control it. Even White misses an essential part of this situation: that the *Homo sapiens* body does not actually go to another scale. Rather, in the view from above, the *Homo sapiens* views another scale but still exists at the meter scale. Just as one can only act on the scale at which one exists (3.34), viewing another scale is not equivalent with intervening in that scale.[15] White falls prey to this conflation when he notes that the view from the airplane allows him to, in some sense, "see the future": "I knew that the car on Route 110 would soon meet up with the other car on Route 37, although the two drivers were not aware of it. If they were about to have an accident, I'd see it, but they wouldn't."[16] The crash here is only imaginary. If he were to communicate with both drivers to tell them of the impending interaction, the drivers would think it absurd: so what if they are heading toward each other? It is their role as drivers to navigate that lower-scale interaction. In translating scalar views to scalar relations, *one must always make use of some intermediaries*—either because you are literally too distant or too large or small to make the intervention directly (see 6.29–35). In such extensions, a far more subtle yet just as essential synecdoche has occurred: you fill in yourself for the whole complex of extensions that make the scalar intervention possible. How can we properly trace out scalar relations if we continually commit this kind of synecdoche?

EXPUNGING THE INTUITIONS OF POWER: THE FEUDAL BETWEEN HOBBES AND WHITMAN

In political theory we can see most clearly how our conception of scalar relations is riddled with nonscalar assumptions. The famous frontispiece of Hobbes's *Leviathan* contains the great representation of the scalar synecdoche in which the king fills in for the aggregate of humans he rules (Figure 9). Clearly, this is a synecdochic scale-trick in which the king comes to fill in for the aggregate. The question is: have we truly thought through a scalar vision of this aggregation that is purged of this basic synecdochic maneuver? And, lest we think this problem only manifests in political and social organization, consider how readily we perform a similar scale-trick on the body: "I" am said to be in control of and responsible for this whole aggregate of cells, molecules, and microbes contained within the boundaries of this skin. This statement restates the central thesis of the previous chapter.

In some ways, the development of democracy was a watershed moment for refusing this synecdoche and rethinking social organization through a more scalar view of aggregation. While there are undoubtedly versions of democratic theory that head in this direction, we can follow Walt Whitman's characterization of the problem as something we are trying to work toward rather than something we have already achieved. I'd place Whitman's *Democratic Vistas* as the foundational text for a truly scalar political theory. He places the task of democracy as a task of surmounting the "history of Feudalism," his term for this set of assumptions in which the aggregate is subordinated to a subset of that aggregate.[17] His text is replete with scalar analogs—the geological (31), the digestive (25), the meteorological (68), the microscope (14), the atomic (30)—as

FIGURE 9: Behold the synecdoche. Frontispiece to Hobbes, *Leviathan*. Andrew Crooke, 1651. Holmes Collection, Rare Book and Special Collection Division, Library of Congress.

he attempts to refigure the task of democracy. He articulates the task as a rethinking of the individual and the aggregate in a way that simultaneously respects the individual and the aggregate without reducing one to the other:

> We shall, it is true, quickly and continually find the origin-idea of the singleness of man, individualism, asserting itself and cropping forth. . . . But the mass or lump character, for imperative reasons, is to be ever carefully weighed, born in mind, and provided for. Only from it, and from its proper regulation and potency, comes the . . . chance of Individualism. The two are contradictory, but our task is to reconcile them. (15)

In scale the two are not contradictory but rather conceivable as two different scales of viewing what humans are and do. Whitman takes us in that direction in his imperative to simply "include all" (21) while situating these individuals in the capacity to rule themselves (22). But his main provocation is that we have yet to discover and fully integrate a scalar view of political, social, or economic organization. He places the vista of democracy at this juncture where we rewrite our whole way of speaking and thinking about our relations in a way that expunges these nonscalar assumptions. If this goal is yet in the horizon, it will be useful to examine a few accounts that head in this direction in order to identify further remnants of a nonscalar view of relations.

CRITIQUES OF (NON)SCALAR DIAGRAMS: SOCIOLOGIES, GEOGRAPHIES, ASSEMBLAGES

Curiously, recent arguments moving us toward such a scalar articulation come from those critiquing scale. These thinkers are resisting the ways erroneous scalar discourses have confused or misplaced the individual and aggregate. We can highlight a few here: Bruno Latour's actor-network theory, the geographer Sallie Marston's critique of scale, Nancy Fraser's *Scales of Justice*, and Manuel DeLanda's assemblage theory. Despite their differences, they share a tendency toward a flattening of relations (in DeLanda's terms, a "flat ontology"), moving toward some conception of a network as a nonhierarchical field of relations that emphasizes how these relations are created and mapped. This map is then said to be inherently about structures of power, exclusion, and totalization. All four are trying to refuse the synecdoche just discussed, but since they are unable to see this synecdoche as a scalar error they treat their articulations as a critique of scale.

Latour's actor-network theory (ANT) treats all things as actors within a network of relations. ANT integrates "nonhumans" into social theory—including scalar objects such as microbes.[18] Each "actant" participates in a field of relations and has a role in the network. This framework pushes against the synecdochic subordination of some objects to others (e.g., the cell subsumed to the body). Furthermore, when discussing social science methods, Latour attempts to avoid the micro and macro distinction, in which a researcher might subordinate local instances to general principles, the reverse, or some compromise between them. He positions this avoidance against the Hobbesian view when he argues that "the body politic is a shadow, a phantom, a fiction" (171).

Such an account includes objects on different scales, but its "irreducibility" creates a problem: how does one treat both cell and body as equally relevant things when they are, in some respects, the same thing? This flattening leads Latour to treat scale as just another relation: "A new topographical relationship becomes visible between the former micro and the former macro. The macro is neither 'above' nor 'below' the interactions, but added to them as another of their connections. . . . There is no other known way to achieve changes in relative scale" (177). Likewise, Latour argues that "scale is what actors achieve by scaling, spacing, and contextualizing each other" (184). Scale, in this view, is another variable and contingent relation. More recently, Latour further claims that "there are not, in the world, beings large or small by birth:

growth and shrinkage depend solely on the circulation of scripts."[19] He suggests, further, that we must "resist the temptation to take differences in size as explanatory in itself."

In this view, scale is produced and, as such, is a contested notion that is reworked and mobilized for various ends. The extensive conversation on scale in geography is largely founded on this basis. The problem is that geography, particularly human and political geography, is primarily interested in mapping human structures of power. As we noted in 3.37, geographers split scales into a largely human series of layers: body, urban, regional, national, global.[20] This schema conflates the logical typing of human relations (discussed in 3.2–9) with the logical typing of scale domains, resulting in a skewed, human-centered view of scale. Doing so produces the critiques of scale as a socially constructed hierarchical ordering of power, since these scales are only discernible within human relations and are clearly a way of mapping those relations.[21] Furthermore, this conversation often extends the philosopher Henri Lefebvre's notion of space as produced to scale in a way that is echoed in Latour's pointing to the "very production of place, size, and scale."[22] Thus, when Marston, Jones, and Woodward call for a "human geography without scale" they point to a "need to expose and denaturalize scale's discursive power."[23] These geographers would discard scale entirely due to the difficulty of "disentangling scalar hierarchies from . . . its army of affiliated binaries" (421). By this they mean the value-laden ideas we have about the local or the global—for example, associating the global with the abstract, open, or detached. Finally, the authors use Haraway's notion of the god trick (see chapter 8 here) to argue that removing scale from geography resituates the researcher at the scale of the body: "How . . . can a researcher write seriously about situated positionality after having just gone global?" (422).[24]

Nancy Fraser's *Scales of Justice* likewise limits scalar distinctions to the human, distinguishing the domestic, the national, and the international as the significant scalar shifts. From these assumptions, Fraser sees scale as a means of navigating who matters, using the idea of the frame or the map to note conflicts of the "who" of any political body. When conceived in this way, politically charged questions of inclusion/exclusion arise (the statement at 3.21 that "scale is inherently inclusive" is responding to this conception). Fraser likewise focuses on scale as socially constructed, speaking of three "frames" by which we organize the scalar relation of politics: "Given a plurality of competing frames for organizing, and resolving justice conflicts, how do we know which scale of justice is truly just?"[25] This framework of frames places justice within a limited conception scale. At the same time, Fraser acknowledges that current conflicts emerge from the breaking down of classic scalar designations, such as the power ascribed to the nation-state. In an appropriate denial of the scalar synecdoche, Fraser points to the undoing of the "political imaginary" that "invested a state with exclusive, undivided sovereignty over its territory."[26]

Manuel DeLanda's assemblage theory similarly critiques totalizing practices by combining the language of systems science with the larger Marxist tradition that has critiqued top-down hierarchical notions of power.[27] DeLanda uses emergent properties and irreducibility (that the parts cannot be reduced to the whole) to create a nuanced concept of assemblages, but his means of justifying this integration reveals some confusing assumptions:

> Without something ensuring the irreducibility of an assemblage, the concept [of the assemblage] would not be able to replace that of a seamless totality. If the parts of a whole are reducible, then they form an aggregate in which the components merely coexist without generating a new entity. . . . Making the properties of a whole depend on the interactions between its parts ensures that these properties are not taken to be *either necessary or transcendent.*[28]

The rejection of the transcendent and the necessary here is another attempt to push against the scalar synecdoche. DeLanda argues that we rework "reified generalities like 'the State'" (16) while still arguing for the use of "levels of scale" to articulate layers of assemblages (17–19). However, his articulation lacks a theory of scale to account for these levels and how they affect assemblages. His rejection of the language of transcendence mixes a critique of scalar synecdoche with the thresholds of affectability implied by scale (3.27–28). Thus, even though DeLanda is right to note that different things can be considered individuals on any given scale, he follows Maturana and Varela (see chapter 8 here) in granting this individuality too much ontological status (22). Most importantly, he suggests that "entities operating at different scales can interact with one another, individual to individual, a possibility that does not exist in a hierarchical ontology" (20). But have you ever actually interacted directly with a single cell in your body qua cell? Have you done so with the ecology of the planet?

In all, we can be sympathetic to these accounts as working toward methods of tracing scalar relations and highlighting some of the problems in articulations of scale. But without a theory of scale, the second interferes with the first. Undoubtedly, a properly scalar assemblage theory, geography, legal-political schema, or ANT can be crafted, but each would require a new articulation and justification that took into account our scalar observations. There are a number of things that we need to account for in particular:

First, we must be wary of nonscalar notions of hierarchy or layers of relations, understanding that they arise from humans attempting to apply nonscalar conceptions to scalar shifts. The diagram of scales based on human criteria is only the most obvious of these (3.16).

Second, scale is not "made" in the same way that assemblages are made, and it is prior to the definition of "actants." Assemblages are assembled.[29] They are formed, reformed, and reworked by various connections and actions. Scale designates levels of observation. Thus, scale is made in a very different way than assemblages and, in some sense, exists apart from them. We cannot define "actants" without this consistency whereby we delimit the sites for tracing relations. Size does, in fact, matter; otherwise you lose reference to what you are talking about, because scalar relations always deal with the same thing viewed on different scales (1.20).

Third, critiques of narratives of scale are not the same as scalar critiques. As a non-arbitrary script establishing a consistent means of accounting for perspective (2.3), scale contains means of separating narratives about scale from what scalar observations demonstrate. Again, this is more potently situated in the conflict that occurs when human-based nonscalar assumptions come into tension with what we find when we scale beyond the human. When Latour, Fraser, and Marston claim that scale is socially constructed, they are conflating these two operations. In addition, when Latour argues that scale is a circulation of scripts that is not invariant, he is confusing the products of scale with the apparatus of scale itself. The products of scale are indeed variant: at one moment we rally around bodies as products of DNA, at another as parts of an ecosystem, at another we tie the two together through multinational corporations manipulating the flows of genes. But scale itself is not variant; if it were, again, we would be unable to track what we are even talking about. Likewise, geographers are right to note that conceptions of scalar action are created, contested, and mobilize for various ends, but this politics of scale inevitably comes into contact with the possibilities and constraints of scale itself.

This is essential for the critique of totalization. Fraser, in particular, is tracing the unraveling of a narrative that is a doubling down of the Hobbesian view of scale. As historian Anne Harrington has shown, there is a direct line between early-twentieth-century nationalism and a form of holism that emerged in science as a response to reductionist views of biology.[30] The emphasis on "the whole as greater than the sum of its parts" provided a basis for a modified synecdoche, combined with the assumption that the larger rules the smaller, in which the nation becomes the reference point that subordinates its subjects. Harrington shows how this frame is then adapted into Nazi notions of wholeness. Fraser then places this narrative in the context of the attempt "to re-engineer human life on a mass scale."[31] These endeavors arise from a tainted, human-exclusive, power-driven view of scale. To the Hobbesian synecdoche they add the scalar errors in which we assume that to unify is to totalize and therefore one must police the who and the how of inclusion. But scale is inherently inclusive within any given domain (3.21), and unity is a function of a change of perspective (1.22), not something made. In critiquing this narrative of scale, as Arendt is doing following World War II, we find ourselves rejecting scale itself, when, to the contrary, more carefully attending to scale helps dismantle the basis for that totalizing worldview.

Fourth, we need to attend to when there is a scale shift versus a logical typing of assemblages or relations. Assemblages can be across scale domains but do not have to be. This is a problem that DeLanda inherits from systems theory, which we will explore in the next section.

Fifth, we need to be clearer about emergence and its relation to both patterns at the higher scale and individuals at the lower scale. That is, these cannot simply be flat relations, since any object in a given scale domain can only act on another scale domain through an aggregation or disaggregation (3.34, 6.30–32). We need to modify our vocabularies of relations so that we do not preserve these assumptions about causality while still tracing out how, for example, bacteria mobilize into organismic effects or humans mobilize into global action. Doing so is difficult, partially because we often want to preserve the agency of objects relevant at one scale into another scale, as we do in speaking of global action. But the question should first be: what *is* the scalar object? Is a corporation, for instance, a scalar entity in the same way our body is? From the scale of the Earth, what does it look like? We need to be on guard for confusions that arise out of these shifts in scale. For instance, we could conceptualize autonomy or agency of actants as both scale-specific (acting on a given scale [3.34]) and arising from scale specificity (something can be treated as a unit because it appears as an object at a given scale [1.19, 5.11]). Indeed, much could be developed out of DeLanda's observation that "larger assemblages emerge as a *statistical result,* or as a collective unintended consequence of intentional action."[32]

Given these problems, we can agree with Whitman that such careful clarifications are on the horizon and that it will take time and significant adjustments to our ways of conceptualizing our relations and our economic and political structures for handling them. Indeed, the provocation set out in 3.23 that ecology encompasses economics, politics, and culture alone requires a seismic shift. In integrating nonhumans into the mix and attempting to think through how to integrate them into human relations, perhaps these thinkers have pushed us in this direction. But this inability to handle scale confuses the terms at the outset. Perhaps this confusion merely shows that we need to more fully expunge the nonscalar intuitions of power. To this end, I want to turn now to systems theory and second-order cybernetics and consider the same questions about scale.

CELL THEORY, SYSTEMS THEORY, AND THE SYSTEM DIALECTIC: BERTALANFFY AND THE TRACING OF SCALAR RELATIONS

Harrington's history of holism in the sciences points to a larger dialectic generated by reductionism between using the small to explain the large versus the large to explain the small. She traces this dialectic through nineteenth-century cell theory, which was framed by Rudolf Virchow in 1855 as evidence that "life was nothing more than the activity of individual cells."[33] Harrington positions cell theory as one moment in an "uneasy balancing act between individualism and unity" that sets up debates about vitalism and holism (12). Thus, Virchow too stated that "the significance of the constituent parts . . . will at all times be found only in the Whole" (12). Considering the cell as a base unit produces both the potential reduction of life to the cell and the framework to consider how these parts come together in an organism.

The relatively late arrival of cell theory represents a crucial moment for scale, because it forced scientists to consider the intermediary points of reference that arise at different levels.[34] With cell theory it becomes newly possible to question how these atoms and chemicals come together.[35] Thus, Virchow positions the cell as a halfway point between the organism and atomic-level action: "Within this locus [of the cell] it is mechanical matter that is active—active according to physical and chemical laws."[36] While this in some way reduces life to the cell and the cell to the molecular, it also makes it possible to see the cell as the intermediate point of consideration. No longer is the question about how atoms make organisms, but first how atoms make single-celled organisms, and then how cells make multicellular organisms.

With cell theory in place, the question of how the "whole is greater than the sum of its parts" enters in the early twentieth century not simply as a vitalism but as a question about how we cross scalar thresholds to form larger-scale entities.[37] The history of this transition to a scalar view remains to be written; here I merely want to highlight how the question of parts and wholes becomes scalar in a new way once we conceptualize these as layers of interaction whereby the relations of atoms become relations of cells that become relations of bodies. This more scalar view makes possible not only the reduction of life to nonlife but the counter-reductionist, larger-scale move whereby one invokes the larger to explain how the smaller comes together. When systems theory is developed in the twentieth century by the biologist Ludwig von Bertalanffy, for instance, he sees his work as firmly in this dialectic, situating systems theory against both analysis of elementary units and mere summation.[38] He argues that biology needs to account for "conditions within the whole" and for "properties that are absent from its isolated parts."[39] To do so, biologists must think of life in terms of systems, organization, and levels in order to provide a new dimension of explanation: the "formulation of principles that are valid for 'systems' in general."[40]

Bertalanffy does not need to conceptualize this move as antithetical to physics even as it refuses the traditional reductionism. To the contrary, he notes that "the advancements of physics tells *[sic]* a different and much more exciting story," where "new fields of phenomena, especially of organization, occurs *[sic]* in the way of a synthesis in which originally separate fields fuse into an integrated realm."[41] This upward movement makes a more fully scalar conception of forces and relations possible. *Until we have this counter to reductionism, we do not have a scalar conception of relations*, since scientists merely go to smaller and smaller scales as the prized domain for interaction. As soon as we work back from the small to the larger, scientists find themselves needing at least three new kinds of principles: *principles of aggregation* to account for how the small gets to the large (6.30), *principles of conditioning* to understand how the large sets conditions for the small (6.31), and *principles of specification* that account for what scale is relevant in any observation (5.11). In introducing these within the notion of the system, Bertalanffy points to what we can call *scale complementarity*:

> There is a kind of complementarity between analytical and global treatment of biological systems. We can either pick out individual processes in the organism and analyze them in terms of physico-chemistry—then the whole, in its enormous complexity, will escape us; or we can state global laws for the biological system as a whole—but then we forsake the physico-chemical determination of the individual processes.[42]

Scale domains produce a scale complementarity: depending on the scale of observation, you will see and interact with different aspects of a system. Given this complementarity, we can then articulate principles and possibilities of aggregation and conditioning (see 6.30–34).

With this properly scalar conception, the dialectic between holism and reductionism can end in the practice of scalar specification. In Bertalanffy and like thinkers we begin to find a more fully scalar view of life where "the unifying principle is that we find organization at all levels."[43] It is for this reason that this book is pushing so adamantly against the anti-hierarchical, anti-transcendent sentiments prevalent in the humanities. Contrary to Latour's suggestion that hierarchies and cybernetics cannot grasp the complexity of relations, without such theories we will be unable to properly trace our mappings and interventions.[44]

INNOVATIONS IN THE TRACINGS OF SCALAR RELATIONS

If scale is central to all of science, then we can query some scientific productions to demonstrate how one might articulate and map scalar relations, but also to identify remnants of the scalar synecdoche and like problems in these articulations. We can apply these questions to the entangled discourses of cybernetics, systems theory, information theory, statistics, computational biology, and symbiogenesis theory. Doing so can highlight possible ways of building things up, but it will also further highlight the persistent remnants of a nonscalar view in these new innovations.

Take, for instance, cybernetics. Cybernetics created a potentially scalar means of mapping how system-level constraints provide grounds for smaller-scale behavior. An account of scalar relations requires formalized tracing of levels of feedback within a system (6.45). The problem is that, from the outset, the most vocal proponent of cybernetics, Norbert Wiener, described feedback in terms of control from a "centralized regulatory apparatus."[45] Doing so carries forward the scalar synecdoche, leading to critiques such as those levied by the historian Peter Galison, that cybernetics was formulated "explicitly in the experiences of war" and preserves, at the core, this desire to control.[46] Yet this synecdochic, controlling view is not as unavoidable as Galison argues. In 1966, Bateson resituates cybernetics in the context of the self-perpetuating feedback loops generated by war, arguing that cybernetics provides some means of reflecting on these loops: "The question is how can we get away from the rules within which we have been operating."[47] He argues that "latent in cybernetics . . . [is] a means of changing our philosophy of control and a means of seeing our own follies in wider perspective."[48] In tracing the structure of feedback and self-regulation within systems, one can see the disparities in our own conceptions of control. In doing so, we might start with the synecdochic interpretations of any central notions from cybernetics (emergence, autopoiesis, feedback, etc.). In what ways can we call intervention in these things control?

A parallel problem arises as cybernetics becomes entangled with systems theory: when is a system a scalar system? Must systems be scalar? Or can we distinguish between systems that remain on one scale and systems that cross scale domains? Bertalanffy, for instance, defines a system as "a set of elements standing in interrelations."[49] But there is a substantial difference between elements on the same scale domain and elements across scales. If the human body is a system, do we mean a system of cells, microbes, and molecules or a system of organs? The latter does not cross a scalar threshold. The problem becomes particularly apparent when sociologist Niklas Luhmann applies second-order cybernetics to sociological systems. The primary examples for Maturana and Varela's autopoiesis (see chapter 8 here) are cells.[50] When we apply the same schema to social systems, however, "systems" such as the nation-state are not a scalar analog to cellular maintenance of molecules.

Maturana and Varela disagreed on this extension of autopoiesis. They provide the basis for autopoietic hierarchies, citing multicellular organisms as the example (110), but then ask:

> What about human societies, are they, as systems of coupled human beings, also biological systems? The answer to this question is not trivial . . . such an answer requires the characterization of the relations which define a society as a unity (a system), and whatever we may say biologically will apply in the domain of human interactions directly, either by use or abuse. (118)

Varela, however, disagrees with this extension into the social, and so the authors go no further there.[51] But in his single-authored introduction to *Autopoiesis and Cognition,* Maturana does go further and describes social systems as autopoietic, giving noticeably human and nonscalar examples: "family, a club, an army, a political party, a religion or a nation" (xxviii). Luhmann follows this extension, with confusing results. This extension grants various kinds of emergent and scalar attributes to structures that do not possess these attributes. Quite simply, claiming that a corporation is an autopoietic system would be equivalent to claiming that the nervous system made and controlled the entire body—another scalar synecdoche.

Thus, we can add to our discussion of autopoiesis (from chapter 8) three observations about the synecdochic nature of this focus on self-reference and self-maintenance. First, these notions are easily mixed up with a kind of scalar colonizing fantasy: that if a group of people claim to be a group ("self-reference") and work to maintain that group (self-regulation), then they are a system.[52] Thus, Luhmann's systems theory emphasizes the synecdochic reference point where organizations "can communicate in their own name."[53] This is exacerbated by, second, the stipulation that a system is independent from external factors—that it is constituted by this division between self and other. Luhmann is emphatic on this point: "The lack of operational access to the environment is a necessary condition for cognition."[54] This leads, third, to the assertion that those entities within the system are contained and subordinated to that system for the sake of its self-maintenance.[55] These points maintain a kind of synecdoche and an overextension of principles of organization on one scale to possibilities and diversity on another. Most importantly, they indulge some terrible habits of nonscalar thought: if I consider the nation or the church as a self-referential system that must maintain itself over and against those outside of it (its environment, those from another country, the nonbelievers) and subordinate its components to this maintenance, then we find ourselves with a biological justification for the kind of othering that claims those considered "components" (of a nation, religion, race, etc.) at the exclusion of everything else. But this is not even how cells work, nor is it how we get from cells to bodies.

At the same time, both cybernetics and systems theory are building on one essential opportunity provided by information theory: one way of explaining the revolutionary aspect of information theory is that it scales communication or semantic structures in a way that was not previously possible. By defining mathematically the conditions of communication within a signal, Claude Shannon provided a different scale view of the process of semiosis. Doing so highlights three further questions about scale that remain in systems theory, cybernetics and subsequent developments:

First, how does information make use of lower-scale components to encode means of communication, that is, imparting forms and patterns that serve as structures to be responded to? Does it make a difference if the ground of a communication is selected on another scale, as in the case of a molecule producing a scent or in a series of transistors encoding for computation? This problem is already implicit in Warren Weaver's separation of three levels of communication: the technical, the semantic, and the effective.[56] Can these be redescribed in terms of scale, perhaps as the lower-scale medium (the technical), the scalar translation (the semantic), and the large-scale effects (the effective)?

Second, how do forms of communication coordinate together smaller-scale entities to produce larger-scale forms? Most systems theories hold that communication is central to the formation of larger-scale human endeavors. Such claims are not so controversial for social communication, because few would question that humans communicate. But can this question of coordination via communication also be extended to cells, bodies, and ecologies?

Third, how does information pass from one scale to another? What, for instance, is information about climate change, and how does that translate to this-scale such that it can actually make a difference? Or, as is the task of the increasingly popular discipline of epigenetics, how do we get from DNA to bodies? An even more provocative way of posing this question is to return it to the questions of experience raised in chapter 8: how do we experience information from one scale when it becomes apparent on another scale? For example, how is the experience of hunger a manifestation of information about another scale (the availability of the appropriate molecules to keep cellular metabolic processes running)? What happens when we have scalar descriptions to make us aware of this scalar transfer? We will make some moves in this direction in chapter 12.

Cybernetics, systems theory, and information theory all inherit additional basic problems arising from statistics. Although not all statistics are scalar, some are, and scale helps us understand something about their operation (see 4.21). Clearly, statistics are essential for mapping scalar relations. Lest we presume that statistics are inherently problematic, we must remember that it is also in such aggregation that we can see structures of inequality and ecological effects. But as we inevitably experience in the frustration of being treated as a demographic, the allure of statistics brings a few essential scalar confusions. In terms of scale, statistics provide descriptions of patterns that emerge in the aggregate while still retaining reference to the smaller components. Therein lies the conceptual difficulty but also rhetorical power: statistics provide a means of redescribing how objects on a smaller scale relate to a larger one. Statistics therefore say less about the actualities of those lower-scale items, although they can provide information about possible ranges of behavior, basis for categorization, and tendencies in those categorizations. The major error is thinking that statistically derived principles translate directly to any single component on the lower scale. Statistics do not annihilate the diversity on the scale below; they identify patterns within that diversity. We must be wary of which scale is being privileged as a mode of description, with the understanding that statistics can point to a larger-scale pattern that must usually be translated to smaller-scale particularities. When such patterns are derived across scalar thresholds, it is important to note that these principles, laws, and trends are not explanatory (6.34); they don't necessarily say how or why behavior has aggregated in the way it has. In fact, it is the beauty of statistics that they don't have to.

We can see these problems in Geoffrey West's *Scale*, which contains many essential descriptions of mathematically derived scalar principles. West, a mathematical biologist, positions mathematical principles as "universal physical and mathematical properties of these networks."[57]

But these laws are universal in a particular way: in abstracting away from any particular object. Such principles "embod[y] a 'universal' quality . . . [by] transcending the specific details of the object that is moving" (77). At times, West acknowledges that this implies that such laws are simply approximations.[58] But at other points he overemphasizes the mathematical principles, stating that such principles are "quantifiable laws that capture" the essential features of living systems (98). This language might lead one to miss the essential qualifier that such principles are "emergent laws of biology" that apply to the "generic coarse-grained behavior of living systems" (98). In other words, there is a scalar maneuver implicit in these principles, and their universal applicability needs to be understood in this moving to a larger scale to explain behavior on a lower scale. Without this point we find a kind of reversed reductionism blended with systems thinking: the supposition that something might be explained entirely by what can be seen in the aggregate. Such a position will always be pushing one scale larger, using the aggregate of cells to explain cells, the aggregate of organisms to explain an organism, and so on.

The great value of West's mathematical scaling is that it traces how "effective laws that govern . . . behavior must have evolved at all scales" (103). In doing so, West allows us to separate out three kinds of scalar principles: those involved in any given object getting smaller or larger (what is known as allometric scaling), principles involved in moving across scales (6.32), and principles and constraints available within any given scale domain (6.49). Statistical methods help us identify some of these attributes as long as we understand when and how we are abstracting from the particulars available at any given scale.

The promises and limits of such mathematical tools have been greatly clarified by computational methods that systematically extend small-scale, localized attributes into larger patterns, such as Benoit Mandelbrot's fractal geometry, Stuart Kauffman's Boolean networks, and Stephen Wolfram's cellular automata. These methods describe and replicate complex patterns of behavior using relatively simple rules or algorithms, using the processing power of computers to compound simple rules in order to create complex behavior.[59] They demonstrate and mimic how simple constraints and tendencies of lower-scale entities create complex behavior in the aggregate. They also imply that one does not need fully descriptive or predictive principles to produce the diversity of the world.

Mandelbrot's fractals, for instance, use simple algorithms to create recursive repetition with sometimes regular, sometimes irregular changes. The resulting patterns are noticeably lifelike (see Figure 10), and Mandelbrot frames his work in terms of creating a geometry of Nature.[60] Many of his examples deal with objects that can be understood as large-scale patterns of lower-scale phenomena including galaxies (84), turbulence (97), cell membranes (114), and Brownian motion (232). Fractal geometry thus describes a kind of aggregation, particularly aggregation that results in repeated patterns across different scales.

Similarly, Kauffman's early Boolean network simulations of evolutionary processes demonstrated the thresholding that can occur in the process of self-organization. These computer simulations of evolution generate "fitness landscapes" that show discernible patterns of change based on simplified variables, models that he describes as a "God's-eye view" that show "surprising large-scale features."[61] From these he derives his concepts of autocatalytic closure and reaction networks, in which a network of components finds itself interconnected in such a way that they begin to function as a unit (50). To illustrate this, he provides a scalar "chemical creation myth." This imagines a series of disconnected buttons that one then begins to tie together.[62] As one creates more connections between the buttons, within a small number of iterations one finds a critical threshold where if you pick up any button, the whole network would be lifted together. Once such crystallization occurs, Kauffman argues, a new level of organization emerges wholesale.[63] He applies these notions to the genesis of life from molecules (61), to the creation of multicellular organisms (126), and to global evolution, economics, and language.

Wolfram's work on cellular automata is equally surprising.[64] Here "cellular" means the cells of a grid that can turn on or off depending on simple rules run many times. In these grids, complex behavior can be created out of a few simple rules and the appropriate starting conditions (see Figure 11). Like fractals, the resulting patterns are eerily lifelike, leading to this line of research being named "artificial life."[65] Wolfram systematically catalogs possible cellular automata behavior, noting the extraordinary and diverse patterns created.[66] He then applies these general principles to a list of scalar aggregates: crystal growth (369), breakages (375; see section 6.38–39 above), fluid flow (376), growth patterns (400), and financial systems (429).

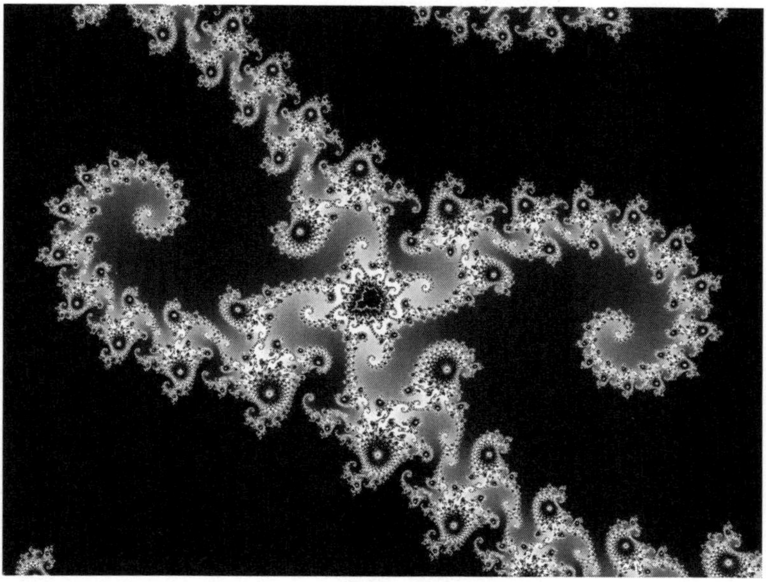

FIGURE 10: A snapshot of the Mandelbrot set. These complex patterns are generated out of a relatively simple algorithm run repeatedly. This image is created by Wolfgang Beyer and licensed under CC BY-SA 3.0.

Mandelbrot's, Kauffman's, and Wolfram's methods rework our intuitive notions of how complexity emerges. Thus, Wolfram argues that his work requires that we develop a "new intuition" (47). If simple rules provide the basis for aggregation into complex forms at a larger scale, then many common assumptions about the means and methods of controlling aggregates are incorrect. These theories suggest that the synecdoche is only possible as a post hoc claim that hardly understands its own maneuver: things and patterns emerge across scalar thresholds, and then one claims (to control, to understand, to be) them without understanding the condition of their emergence. We then seek the rule that will produce or reproduce a behavior, but the rules provided here are not controllable in the traditional sense. Thus Wolfram argues that even though such models "capture the essential features of systems with very complex behavior" (750), they are "computationally irreducible—so that in effect the only way to find their behavior is to face each of their steps" (6). In other words, "the only way to find out what they do will usually be just to run them" (750).[67] Similarly, Kauffman

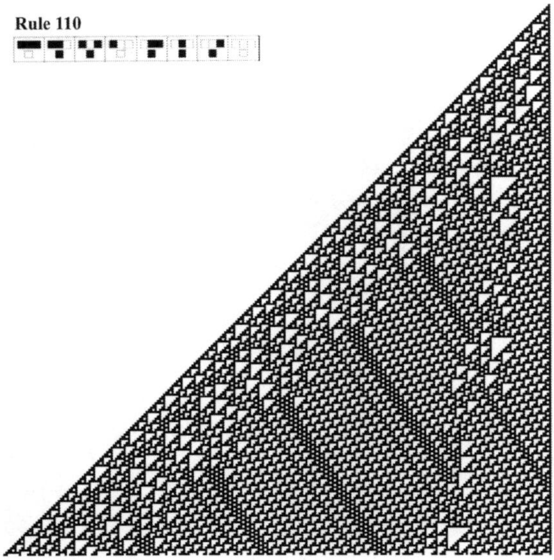

FIGURE 11: An example of Wolfram's cellular automata. These start with a single line (the top right of the diagram). One then sets rules for what will happen in the next rows, using the three previous cells above any given cell to determine whether this next cell will be black or white (this rule is pictured in the top left). This produces 256 possibilities in the simplest version, most of which are simple or repetitive, but some of which are surprisingly complex. This is Rule 110 run to 250 steps, which demonstrates regular yet complex patterns. Others, most famously Rule 30, produce seemingly chaotic non-periodic behavior. Image copyright by Stephen Wolfram, LLC.

argues that "what we are after here is not necessarily detailed prediction, but explanation" (23). But it's a specific kind of explanation, a theory of emergence that "gives up the dream of predicting the details" (19).[68]

What these methods seek is an understanding of the possible forms and constraints of scalar aggregation (6.30–32). For instance, consider the notion of attractors, which Kauffman describes as the area that a pattern tends to converge on (78–79).[69] Kauffman notes that "under the right conditions, these attractors can be the source of order in large dynamical systems" (79). "What we need," he suggests, "are laws describing which kinds of networks are likely to be orderly and which are likely to succumb to chaos" (79). These models of aggregation are attempts to understand conditions and possibilities of the emergence of particular changes across scales. These computational models imply that thinking in terms of growth and cultivation may be more appropriate for all scalar developments, since we allow these aggregations to unfold rather than directly craft them, as we do with many this-scale objects.

At the same time, the approaches just discussed more or less empty out the content of the lower-scale entities in the search for laws of aggregation or conditioning apart from the components involved. This imbalance points to a need for additional theoretical tools for examining the grounds for lower-scale aggregation, particularly in living systems. After all, forms of interaction, particularly cooperation, make possible more pointed aggregation of effects on a larger scale. This is the significance of the work of Lynn Margulis, Jan Sapp, Soren Sonea, and other biologists who pioneered a more properly scalar way of thinking of the collaborative nature of life by considering bacteria more carefully and generously. Margulis, in particular, made two scientific contributions arising from this microbial thinking: her reintroduction, expansion, and defense of symbiogenesis and her early contributions to Gaia theory.[70] As an evolutionary term, *symbiogenesis* refers to "the origin of new tissues, organs, organisms—even new species—by establishment of long-term permanent symbiosis."[71] Margulis also collaborated early with James Lovelock on the Gaia hypothesis, largely because Lovelock needed some means of explaining what on Earth produced feedback mechanisms.[72] Margulis's answer: microbes.[73] Later in life, Margulis reflects on the connection between these two ideas with the phrase "Gaia is just symbiosis as seen from space."[74]

Both of these ideas are exercises in scalar thinking that demonstrate how scalar discoveries rework common assumptions about life, control, and humanity. Thinking with bacteria permits Margulis and her son and coauthor, Dorion Sagan, to provide powerful descriptions overturning humanist, nonscalar assumptions. For instance, in *Microcosmos* they write: "It is not too whimsical to say that if we feel at loose ends, of two minds, beside ourselves, going to pieces, or not together, it probably because we are. Real organisms are like cities."[75] By this they mean not a Hobbesian body politic but a city full of all kinds of people, animals, plants, objects, and, especially, microbes. Symbiosis is the common and pervasive element. The lives of bacteria are already entangled together and constantly interlocking, cooperating, and mashing into each other.[76] These *Homo sapiens* too can be considered aggregates of microorganisms. As the sociologist Myra Hird puts it after an extended study with Margulis, we "must also somehow recognize that 'I' am bacteria, that bacteria are us."[77]

Symbiosis also permits us to set aside two of the most persistent ideas associated with scale: the notion of the hive mind and the "great chain of being." Clearly, the trope of the hive mind preserves a synecdochic view of aggregate cooperation, as if components must be of "one mind" to coordinate behavior. To the contrary, a scalar view of symbiosis suggests that smaller-scale behavior (even noncooperative behavior) can actually produce coherent larger-scale behaviors—even cogitation—at a new scalar level. Indeed, this is already inherent in Teilhard de Chardin's and Vladamir Vernadsky's notion of the nöosphere, the name they give to the mental formations emergent from the aggregate rather than subordinating this aggregate.[78] While Teilhard positions man as an essential moment in the evolution of the nöosphere, this scalar conception does not maintain the classic "great chain of being."[79] In this new hierarchy of being it is not minerals, plants, animals, humanity, angels, and gods arrayed in levels of power, but atoms, microbes, organisms, ecologies, solar orbits, and galaxies as perspectives on the dynamism of the cosmos.

REELING AT/IN THE EXTENSIONS (OF MAN): FANTASIES OF CONTROL IN NANOTECHNOLOGY

Of course, humans do find themselves interacting with different scales in more and more deliberate ways. The problem is that, in our increasing awareness of these entanglements, the category of the human tends to claim these relationships in both affirmative and negative ways. While the last section considered scientific methods for building things up, scale points to a broader need to reconceptualize technological interventions as symbiotic extensions with agency distributed along the apparatus.

While we can feel the distress of this synecdochic claim in ecological crisis, we can see the way it operates in the discourse around nanotechnology. Two texts lay the foundation for the rhetoric of nanotechnology: Richard Feynman's 1960 talk "There Is Plenty of Room at the Bottom" and K. Eric Drexler's *Engines of Creation*.[80] Drexler's influential vision of nanotechnology combines two ways of imagining the scalar intervention. The first, derived from Feynman, suggests that one can make a set of "slave 'hands' at one-fourth scale" and then "wire them to my original levers so they each do exactly the same thing at the same time in parallel."[81] One uses these hands to make smaller hands and attach those hands to this set—and so on until one arrives at the nanoscale so that one might precisely position atomic components. This scheme imagines a synecdochic extension of the *Homo sapiens* to the atomic scale. Feynman thus declares that we could "arrange the atoms the way we want; the very *atoms*, all the way down!"[82] Drexler follows this maneuver: "our ability to arrange atoms lies at the foundation of technology . . . [but] with our present technology, we are forced to handle atoms in unruly herds."[83] In contrast, nanotechnology is a "molecular technology" that will "handle individual atoms and molecules with control and precision" (4).

Yet, Drexler adds to this his most persuasive maneuver: the claim that nanotechnology already exists in cells and microorganisms: "More complex patterns make up the active nanomachines of living cells" (5). This act of renaming cells as "nanomachines" and cellular components, such as ribosomes, as "molecular assemblers," permits Drexler to, in a sense, "prove" that precise atomic manipulation is possible: "Ribosomes are proof that nanomachines built of protein and RNA can be programmed to build complex molecules" (8). This maneuver, which became standard in nanotechnology discourse, conflates not only the biological and mechanical but also two different scales of machinery: "All machines use clumps of atoms as parts. Protein machines simply use very small clumps" (6). At stake is the possibility of extension of human control and interests into the molecular scale. But is there not a significant difference between Feynman's iterated extension of the large scale into the smaller scales and the operation of cells?

Fifteen years later, Richard Smalley challenged Drexler's ideas on this basis.[84] Smalley contends that there are challenges specific to the nanoscale that make this Feynman-extension impossible. The resulting debate between Drexler and Smalley is largely about this tension between Feynman's model and the biological.[85] Smalley challenges the notion of "molecular assemblers" from the position of biology, pointing to the need for water, the messiness of a cellular environment, the need for nutrients, and so forth (40). Drexler's response is surprising; he insists that molecular manufacturing is not biological at all. Rather, it uses "productive machinery—factories—to build smaller factories, leading ultimately to nanomachines building atomically precise objects. . . . Although inspired by biology . . . Feynman's vision of nanotechnology is fundamentally mechanical, not biological" (40). This description reinserts the scalar conflation, suggesting that we can manufacture on the molecular scale in a relatively similar way to the way we manufacture on this scale. Smalley's final response returns again to this biological-mechanical switch, making clear that it is also an assumption about scale and control: "You are still in a pretend world where atoms go where you want because your computer program directs them to go there" (42). Smalley's challenge is about appropriate understanding of scale and the problems that arise in extending this-scale understanding to the nanoscale. As we do find ways to develop machines on the nanoscale, they will need to be able to perceive, make use of, and interact with a different field of interaction that must rely on scale-specific capacities to respond to the dynamism at that scale.

Smalley's point is also about popular understanding of these technologies. In conflating biological-mechanical, this-scale and nanoscale manufacturing possibilities, Drexler must enter a "pretend world."[86] As Colin Milburn has described extensively in his book *Mondo Nano,* nanotechnology has developed largely through virtual simulations, speculative capital projects, simulated interfaces for imaging atomic-level manipulations, and gaming cultures that imagine such interventions. These simulations reflect an ongoing obfuscation of the way that nanotechnology is tapping into attributes already present at the nanoscale. Milburn describes this confusion as the treating of "bits as binary digits, and bits as matter broken apart." But even Milburn seems a little too engrossed in the virtual extension, suggesting that this "dream of digital matter" permits us to "enact our prehension of the molecular world, reaching down to feel its vibrations and tensions, biting into it and getting a taste for it. The . . . repetitive pattern of total nanodissections [in nanotech video games] . . . serves to allegorize the processes—both technical and social—whereby matter is made data and data is made flesh."[87] But we have to wonder: is this digital extension really making matter manipulable or is this a way of narrating the extension—a fantasy contingent on the digital simulation that disguises the conditions of such extension?

Clearly, nanotechnologists will have to learn to work with the dynamics of the nanoscale via more and more complex intermediaries. Yet, even as we move closer to creating nanomachines similar to Drexler's assemblers, we continue to write over these extensions by eliding the implication of the biological conflation. Thus, in a 2017 *Nature* article outlining a design for a molecular machine, we find the same trope again:

> This first-generation machine augurs well for the development of small-molecule robots that can be programmed to manipulate substrates to control synthesis in a form of mechanosynthesis (that is, the use of

> mechanical constraints to direct reactive molecules to specific molecular sites [here they cite Drexler]), in a manner reminiscent of the molecular construction carried out in biology.[88]

Clearly, we are on our way toward some extraordinary and newly reworked entanglements with the nanoscale. But, as we do so, will we continue to elide the dynamism of that very intervention, overwriting in terms of control, manipulation, and direction the highly complex processes that we find ourselves examining and intervening in?

In response, we do not necessarily need to reel in our technological capacities or turn back the clock on the forward momentum of a research agenda that already seems to have its own self-sustaining cybernetic reinforcement. Instead, this discussion suggests the need to reel in our own illusions of grandeur and stop reeling at the entanglements we find ourselves engaged in. Perhaps then we will be able to work with these new scalar relations, avoiding both the fantasy of domination and the horror at the interventions, and come to appreciate these scalar extensions as new forms of scalar symbiosis.

ITERATING THE SCALAR CALIBRATION

Circling around these scalar relations—ecological entanglements, DNA unfoldings, nanotechnological interventions, fractal thermodynamic patterning, cellular aggregations—we find ourselves losing our grip on the familiar and the assumed. Like Schweickart, we look for home, but find, when it circles back around, that home is not what we had presumed it to be. When and if we stop pitching a fit about this new dynamism, we can look around and observe that all the things we hold valuable, essential, and inevitable are reworked in a complex layering of relations.

But the "when and if" is essential, particularly in facing our current ecological imbalance. If we are to "stay with the trouble" and learn to make kin, as Haraway has encouraged, we need to work with and learn to use the rhetorical technologies, as Richard Doyle calls them, that serve as "attention attractors" that manifest and make possible a new responsiveness to these scalar relations.[89] As Doyle argues, the persuasiveness of scalar views "seem[s] to hinge on an *experience* of this interconnection as well as an *understanding* of it."[90] In Part III we will turn to the rhetorical technologies associated with scale, but here we can conclude our discussion of scalar relations by considering the effect of this new configuration: what is your own relation to these scalar relations?

The result, as we saw already in Schweickart, is a recognition of the contingent nature of the categorizations and mappings derived from human experience or imparted by human inquiry. All human values, languages, and conceptions are maps built of the relations and orders of the cosmos, maps that solidify into forms of preference, structures of association, and weighty implications. Scale requires that we bring these structures into view so that we can rework them by hooking them up into relations far exceeding ourselves.

For the scientist, some mode of observing one's preferences is required to achieve something like an objective view in examining these relations as one distances oneself from the categories and relations you think you know. For the nanoscientist, the understanding of nanoscale interventions as not tightly controlled is not some New Age rewriting; it is an attempt to more adequately acknowledge the dynamics of the apparatus. For the biologist studying evolution, symbiosis is not a new Lamarckian trend but an acknowledgment of the variety of factors that contribute to the dynamics of life. For the ecologist, testifying to the role of humans in affecting this planetary ecosystem is not an attempt to insert politics into science but rather an attempt to trace our deep interconnection with the planet. In understanding these relations, values do not have to be left aside; they need to come into view to be reworked. As our value systems inevitably come into friction with these scalar tracings of science, we all need to find means of permitting our values to be transformed by these discoveries.

This task of transformation is not delivered by scientists as keepers of truth, nor is it displaced from us in some kind of social-political seismic transformation. Societies are not the scale at which such revaluation occurs for you—to suggest so is already to drift toward the inverted synecdoche in which society determines value structures within all individuals. Instead, such reconfiguring requires each of us to bring into view the structures of association and value embedded within these brains and bodies. Thus we circle back to the trope of the transformation created by the scalar view, combined now with the trope that one should, first and foremost, tend to oneself. Indeed, this is the trope already in Plato's *Alcibiades*, where Socrates demands that the young ruler first learn to know himself before attempting to rule others. Or likewise, in Socrates's last advice in the *Phaedo* that "If you take care of yourselves you will serve me and mine and yourselves, whatever you do . . . but if you neglect yourselves . . . you will accomplish nothing" (115b). In this task of contemplation, the result is a transformation of your base values as they are brought into view and reworked.[91] As Aurobindo describes it: "By dwelling in this cosmic consciousness our whole experience and valuation of everything in the universe will be radically changed."[92]

This refiguration is already at the core of Pseudo-Dionysius's coining of the term "hierarchy": "The goal of a hierarchy . . . is to enable beings to be as like as possible to God and to be one with him."[93] Such hierarchy is "a state of understanding and activity of approximating as closely as possible to the divine" (153). In other words, before hierarchy was co-opted into a description of human social castes, it was a description of the attempt to reconcile these layers of being and calibrate them to the All. This truly scalar hierarchy does not subsume cells to bodies or bodies to cells—or any similar scalar subordination—but rather "ensures that when its members have received this full and divine splendor they can pass on this light generously and in accordance with God's will to beings further down the scale" (154). In other words, the scalar hierarchy describes and guides these *Homo sapiens* toward the capacity to respond to this new configuration of relations.

We can hesitate for a moment on the zealous tone of "pass on this light," but only to understand that this is not a merely religious thought. Undoubtedly, in the aggregate we can help bring into view this structure of values, but we always need to be wary of misplacing the scale of this transformation. This tension is brought into relief by religion and any other social structure or personal vendetta that attempts to synecdochically claim this process of value regulation for itself. Further tension is brought into view by our current political discourse, as we persistently submit to the scalar confusion where the larger is the more powerful and we permit our leaders to fill in for the whole both in name and practice. As Aldous Huxley wrote in 1945 at the end of the war partially built on this terrifying holism, "the cult of unity on the political level is only an idolatrous *ersatz* for the genuine religion of unity on the personal and spiritual levels."[94]

We can take this one step further and suggest that the tendency today, both in popular forms and in critical theory, to read everything in reference to the political risks something of a synecdoche and scalar rewriting. The personal cannot always be read in the view of the political. Undoubtedly, the personal is political, just as it can also be seen as economic and ecological. But these scalar reference points risk overtaking and rewriting this-scale experience without appropriately tracing the scalar feedback on which they are grounded. Such communal frames might then be configured on competing grounds, as ways of rereading your and, most importantly (as we turn in judgment to our neighbor), others' actions in terms of the economic-political (as in neoliberal views), the ecological-political (in environmentalism), or the ethical-political (generalized standards of behavior, derived from religion or elsewhere, that one would like to impose on the general populace). In each case, one's immediate world and this-scale social interactions are rewritten from a perspective of the aggregate.

Thus, returning to Arendt, we must rediscover that the active life, however important, cannot subordinate or forget the contemplative life. In treating scalar views as an alienation of this-scale existence, Arendt would have us think that such revaluations are antithetical to a proper tuning of ourselves to each other. Thus she reinforces the idea that "the philosopher's experience of the eternal . . . can occur only outside the realm of human affairs and outside the plurality of men."[95] Although Arendt by no means leaves behind the contemplative life, this assumption—that the contemplative is antithetical to the active—is too easily invoked to critique the contemplative.[96] But losing sight of this contemplative transformation is just as much an alienation from this-scale existence, as we disavow our personal transformation in view of the transformation (or lack thereof) in the community. Why does it matter, we say, if I tune to these ecological relations, examine my implicit racial bias, or understand my technologies as symbiotic if we, in the aggregate, do not? The very form of this question is a scalar confusion.

We are currently in the midst of an immense scalar calibration on dozens of fronts, both individually and collectively, as we adjust to these newly elaborate scalar entanglements. As Bateson describes it, calibration is the repeated adjustment of a system in response to feedback.[97] In prodding scalar relations, we are receiving feedback that indicates something about our relationship to these levels of relations. But no one feedback loop is a calibration. Rather, calibration occurs through repeating that feedback so that adjustments can be made based on that repeated information. This calibration is an inevitably difficult process that also occurs on multiple scales. In this feedback we need to remember that even widespread resistance to this information *is itself part of the calibration*. This tension arises because these cross-scalar relations create what Bateson calls a "double-bind," which he describes as the relation between two levels of abstraction improperly calibrated.[98] This double bind arises from the human, at this scale, seeing all these scales of relations that they both are and are not. This double bind is resolved by going to a higher logical level, which is not here a higher scale but to a comprehension of scale itself.

From this perspective, perhaps, we can find ourselves in a new relation that delights in these multiform relations. In this spirit, we can leave off here with Whitman's scalar invocation in which all relations are recognized within the fractal unfolding of what is:

> I believe a leaf of grass is no less than the journey-work of the stars,
> And the pismire is equally perfect, and a grain of sand, and the egg of
>   the wren,
> And the tree-toad is a chef-d'oeuvre for the highest,
> And the running blackberry would adorn the parlors of heaven,
> And the narrowest hinge in my hand puts to scorn all machinery
> . . . .
> I think I could turn and live with animals, they are so placid and
>   self-contain'd,
> I stand and look at them long and long.
>
> They do not whine about their condition,
> They do not lie awake in the dark and weep for their sins,
> They do not make me sick discussing their duty to God,
> Not one is dissatisfied, not one is demented with the mania of
>   owning things,
> Not one kneels to another, nor to his kind that lived thousands of
>   years ago,
> Not one is respectable or unhappy over the whole earth.
>
> So they show their relations to me and I accept them[99]

PART III

Rhetorical Technologies for a Theory of Scale

The Algorithms and configurations presented in Parts I and II do not fully account for one of the strangest aspects of scale: the problem it poses for representation. Because scale occurs outside of human experience, it must be re-presented at this-scale. Inevitably, its representations will be partial and distorted. These representations take a part of this scale and have it stand in for something on another scale. In this process, even science finds itself limited to particular maneuvers that make its descriptions possible as an attempt to track and account for these distortions and piece together these partial views. Chapter 10 will thus examine the operations of science as well as the role of those in the humanities who study science (what is known as science studies). In turn, because scalar depictions will be in some way obscure, they require a particular approach to reading and critiquing that, in a few important ways, run counter to many of the tendencies currently in vogue in cultural criticism. Chapter 11 will thus examine the structure of scalar representations in order to outline ways of approaching such depictions.

In addition, as I have made clear throughout this work, scale provides a transformation not only of reality and but also ourselves. How can these representations produce this transformation? Such representations must not only act as a means of conveying scalar information (what we usually speak of as facts) but also function as rhetorical technologies for assisting us in comprehending scale. All of the chapters here will deal with this question, but the final chapter takes it on most directly, thinking about the scalar relationship between brain and experience in relation to contemplative practices.

Mapping the Vast Unknowing

The Science of Scale, the Scale of Science

> THEAETETUS: What do you mean, everything?
> VISITOR: You don't understand the first thing I say! Seemingly you don't understand everything!
>
> —Plato

> If all knowledge were intuitive, there would be no need for science.
>
> —Joel R. Primack and Nancy Ellen Abrams, *The View from the Center of the Universe*

SCIENCE BETWEEN SUSPICION AND SKEPTICISM

At the edge of the solar system, the disembodied voice of Carl Sagan speaks of the pale blue dot of the Earth and, with that voice soaked with homey wisdom, demands that we take into account the vastness. Television radio waves surpass the *Voyager* spacecraft in which the golden disc soars out into space, awaiting discovery. Within some of these waves lie a thousand iterations of Sagan's opening lines of *Cosmos*: "The cosmos is all that is or ever was or ever will be."[1]

At the edge of the observable universe, Neil deGrasse Tyson takes his place next to his idol to take us once again through the cosmic calendar, this time enhanced by CGI, an even bigger budget, and a major production studio. Yet before Tyson can speak, this 2014 version of *Cosmos* replays the opening clip from Sagan's original performance, repeating this dramatic statement of wondrous science.

And yet, this opening line remains a strange statement. How do we comprehend a science of the Cosmos if it is defined in this way? To what do these words refer? We exist here, in this room, with our everyday affairs always in front of us, occupying our time and attention. All of this, says Sagan, is part of the Cosmos. But so is everything else, including what we haven't yet thought to include. What, then, is the Cosmos that we are called to comprehend?

In Part I we spoke of the Cosmos as an inclusion of all that might be observed on any given scale (3.19) and as all that might be observed at all scales (3.41). The "all" in Sagan's definition of Cosmos aligns his mode of description with the All as described in 3.38: the simple injunction to "include everything." Sagan's *Cosmos* points to how science has continued to expand this All, fleshing it out with quarks and galactic bodies, cells and planetary relations (3.42). Yet this expansion into "all" things is puzzling, since it must include all kinds of difficult-to-comprehend objects, including those we do not yet have the means to understand. This is the topic we want to take up in this chapter. I want to consider what kind of map science is, such that it might hold authority to extend our knowledge into these scales that exceed our immediate experience.

In Sagan's sincerity we find these questions of scientific authority tied to wonder and an effusion of affect:

> Our contemplations of the cosmos stir us. There is a tingling in the spine, a catch in the voice, a faint sensation as if a distant memory of falling from a great height. We know we are approaching the grandest of mysteries. The size and age of the cosmos are beyond ordinary human understanding.[2]

Sagan invokes scales of time and space far beyond the human as the fount from which the mysteries of the Cosmos move us to a new form of contemplation. But, what exactly does this affect-laden characterization have to do with science? Has Sagan lost his scientific grounding? Perhaps we ought to ascribe to Sagan a different role, the prophet, as Ann Druyan seems to do in Sagan's posthumously published *Varieties of Scientific Experience*. Comparing Sagan to the biblical Joshua, Druyan declares in her introduction that Sagan tried to knock down the "wall[s] around our souls that keep us from taking the revelations of science to heart. . . . We are spiritually and culturally paralyzed, unable to face the vastness, to embrace our lack of centrality and find our actual place in the fabric of nature."[3] Situated this way, Sagan's rhetoric serves a moral and quasi-religious function, delivering the discoveries forged within the citadel of science to those bereft of its knowledge.

Of course, there is something obnoxious about this characterization. It reeks of the suspicion that, even as scientists counterpose themselves to religion, they assume its mantle as the arbiters of truth and morality. In response, the scientist rails against this attitude, turning the conversation back to the practices of science, arguing for its rigor, skepticism, and discipline as the ground for this authority. After all, science has long been in the business of resisting the pseudoscientific, the superstitious, and the speculative. Sagan was also adamant in his defense of science and in working against superstition. However spiritual in tone Sagan's work might be, much of *Varieties* focuses on dismissing and otherwise arguing against traditional conceptions of religion and spirituality, and his final book, *The Demon-Haunted World,* is another defense of science against superstition.

The year Sagan died—1996—also brought the Sokal hoax, in which the physicist Alan Sokal published an article in the postmodern theory journal *Social Text* that he then declared a deliberate hoax. If Sagan had been alive to see the years of debate and scornful dismissals that followed, he likely would have agreed more with Sokal than those who defended critical theory.[4] When I began my education in science studies twelve years later, the Sokal affair was still simmering enough that it was taught in my literature and science course. Nothing seemed resolved at all between these two cultures.[5] If anything, the situation has been amplified by another major development: an increase in the tenacity and visibility of overt denialism and obfuscation of science that has now been called the "post-truth" era, which seems to accept a politicization of science in order to undermine disagreeable science. This so disturbed Bruno Latour, one of the targets of the Sokal affair, that his 2004 article "Why Has Critique Run out of Steam?" revisits his role in this controversy, recasting his intervention into science that became the core of science studies.

Sagan, it seems, left this world just when the tensions were about to burst over the epistemological validity of science, the appropriate domain of its descriptions, and the ways to justify its practice and authority. In miming the caustic comparison of Sagan and religion, I want to return us to a moment before the conversation reached this degree of furor. In this context, I want to take some space to comment on both the practice of science and science studies. After all, scale thrusts us across these two cultures and forces us to ask questions anew about these attitudes that recur in both science and cultural criticism: critical, skeptical, suspicious, incredulous, concerned, distressed, ignorant, beguiled, nonsensical. Quietly, late to the party, unfashionably dressed, a ghost of Sagan inserts: wonder? Perhaps there is a mutual bewilderment that can help us clarify the operations of science and thwart this antagonism between the skepticism of science and the suspicion of critique.

Look again at the premise of this book: in scale we describe, encounter, and intervene in existence on levels outside of normal human experience. How is this possible? What are these cells and ecologies? How does science so confidently make sense of them? How can we speak of them in our everyday lives as if we understand what they are? Are we, scientists, humanists, or anyone, so certain that we know what we are talking about—not in the authority sense (as in "I am knowledgeable in the field of microbiology") but in an experiential sense ("I know this body as a menagerie of these semiautonomous bubbles of life coordinated in an intricate aggregation")? Surely there is more to be said to understand a cell than what is written in a textbook and refined, by whatever process, in labs and professional journals. Indeed, the wondrous nature of this transformation implies that this scientific process is not so separate from this wonder. How does this affect relate to the messy yet highly regulated accumulations of scientific practice?

I want to show how the problem of scalar extension and the question of the scale of knowledge gives us one way to resituate the debate over the epistemological status and authority of science. The examples that are most readily invoked as needing the authority of science, such as climate change, evolution, toxicity, or treatments for disease, are usually those that force us to change our behavior in relation to discoveries made at other scales. In turn, scalar examples are those most readily amenable to strong critiques usually called "constructivist" or "relativist"—for instance, Latour's declaration that microbes did not exist before Pasteur.[6] We can reevaluate such statements if we situate them in the particular challenges presented by scale.

The argument here is simple: scale points to how science encodes a not-knowing within a knowing (3.43). Given this skepticism and tension over science, some work must be done to situate this rather mystical claim. To this end, I want to pose a simple contrast of four kinds of description of the same object. This contrast will set the stage for a discussion of how science functions as a regulated process of specification. We can then consider this specification process in relation to the epistemological critiques from science studies. In this context, we will discover bewilderment and wonder as the starting point for discovering and working with the inherent not-knowing within science's knowing. We can clarify this relationship between wonder and science by examining the precedent set by Alexander von Humboldt's description of a "science of the Cosmos." This notion of science will leave room to consider what traditions handle the question of transformation beyond the development of this kind of knowledge (*epistēmē*). To help pinpoint precedents for this endeavor, I will briefly examine this question in Plato's concern about rhetoric and his relationship to mysticism. Doing so will outline the conditions required to follow out and integrate the scalar extensions of science into this-scale experience.

This chapter also hopes to make clear how a concept like scale sits between the sciences and the humanities. Sagan's rhetoric serves as a kind of litmus test for how well we have traced and mapped these modes of description: do we understand where and how to situate Sagan's science in relation to his turn to wonder? The task here will not be a blending of the two so much as a clearer delineation of each so that we can see the function, promise, and limits of both kinds of practice and discourse. Scale will emerge in the middle, operating with the same principles but moving humans to different but equally important and legitimate ends, depending on how it is situated and what it is used for.

FOUR DESCRIPTIONS OF A CELL

In cell theory we find a simple object we can use to consider the operations of science. This body is made up of cells, which operate in partial independence in a different phenomenal world that nonetheless constitutes the grounds for this body. While many object to evolutionary timescales, dispute over toxicity of errant molecules, struggle to wrap their minds around pandemic-level viruses, or fail to comprehend their relation to a planetary ecology, few seem to find the idea of cells objectionable. No one is spending time on the pulpit denouncing the notion of the cell, and no great education is needed to imagine them. Yet I'd argue that everything discussed in this book can be unfolded from dwelling with this significant scalar shift.

We can outline four options for making-present, describing, and analyzing this scalar object. Undoubtedly, one could think of more, but this sketch will suffice. For all but the last, we can use an image to make it paradigmatic. The first description of a cell is an image, produced by

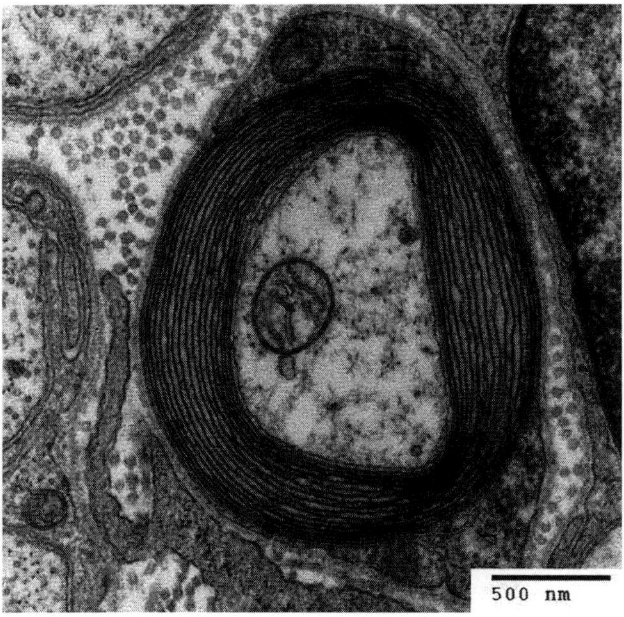

FIGURE 12: First description of a cell: images of—or, more generally, data produced from encounters with—the object in question. This is a transmission electron micrograph of a portion of a neuron. Generated and deposited into the public domain by the Electron Microscopy Facility at Trinity College.

a microscope, which becomes the data on which knowledge of a cell is crafted (Figure 12). Microscopy is the only way we know about cells at all; without microscopes, we could not even speak of their existence. Of course, microscopy was not *sufficient* for the understanding of cells, but it provided the basis for their definition as well as the primary evidence.[7] Such images help form the cell as an "epistemic thing," as Hans-Jörg Rheinberger calls them: "material entities or processes . . . that constitute the object of inquiry."[8] Scientists then develop additional ways of producing data and images used as evidence for further descriptions of this sort.[9]

Out of the process of prodding and experimentation used to isolate and address particular questions about the cell, scientists then produce the next two kinds of descriptions (Figures 13 and 14). These kinds of diagrams and images are essential for conceptualizing the operations of cells, mapping their parts, and tracing their processes, as well as then communicating these results both to other scientists and to

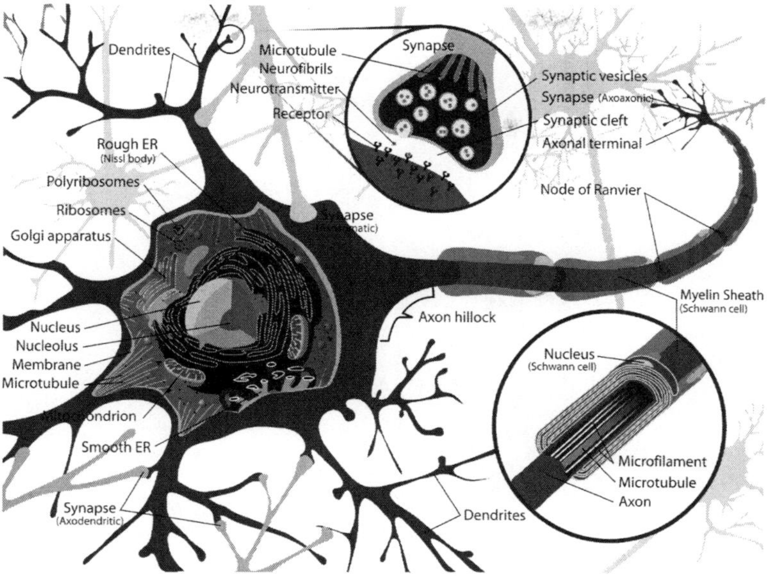

FIGURE 13: Second description of a cell: the diagram of the object. This is a chart showing the parts of the neuron in a way that clearly positions and makes visible all of their essential features.

nonscientists.[10] One need go no further than the famous model of DNA by Watson and Crick to understand the role of these models in attempting to conceptualize objects and processes on another scale.

The fourth description of a cell is quite different. You, the reader, must provide it: it is the fact that these diagrams and image just provided are about your body. This body called "you" is made up of around 37.2 trillion cells, and the brain processing all of this is made up of around 86.1 billion neurons. Your body itself is a description of cells, and your cognitive processing a description of neurons, albeit one that translates from the micrometer scale to the meter scale. What are cells here, in this hand, in these muscles, in this brain?

The first kind of description is any means of examining the world on another scale. It thus also includes other, less direct and less clearly visual modes of prodding scales, such as electron-density maps in protein crystallography or the statistical readouts produced by colliders.[11] These scalar mediations make use of an apparatus capable of cross-scalar translation, tracing differences only discernible at another scale and presenting them for deciphering. The second description is a way

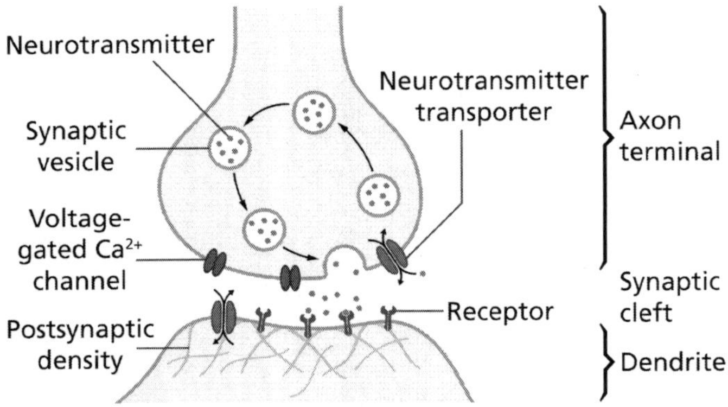

FIGURE 14: Third description of a cell: the diagram of a process. This image charts the transmission of a signal across the synapse of neurons. Synapse Schematic by Thomas Splettstoesser is licensed under CC BY-SA 4.0.

of tracking, placing together, and visualizing this information so that we can make a relevant picture of existence viewed at that scale. The third then mobilizes this picture into a series of relations. All three are essential for the epistemic processes of science; the scalar prodding provides the potential differences, while the models and charts piece together and interpret those scalar encounters. We can then add this fourth description as an essential step: that of bringing the scalar description back into *this scale,* here, and considering its relation to experience.

Should we say that this fourth description is not of cells at all? If so, then we deny any validity and applicability of any scalar inquiries—one would have to say that DNA, climate change, galaxies, viruses, and quarks also have no place in this-scale reality. In this light, this fourth description must be seen as equally important and valid. What are microbiologists describing if not the things in this body? The first three images do not describe the precise nature and activity of the cell in the muscle on your left little finger about halfway between your knuckle and nail, nor do they pinpoint the particular group of neurons involved in processing this sentence. Yet, at least in principle, that is what these descriptions are about. We now can wonder what science is performing: here the most rigorous and sustained inquiry into the nature of these cells appears strangely removed from the reality it is describing. If we are to accept some "reality" in the operations of science, this difference must be the first bridge that is crossed.

A SCHEMA OF SPECIFICATIONS

These four descriptions of a cell bring into view different ways that we can consider the operations of science. To admit the validity of the fourth description is to simply note that there are ways of specifying the domain and operations of a description, practice, and concept, that is, what we usually call knowledge. Scale requires such specification, at the least, because what is described in the aggregate (e.g., neurons) may or may not translate directly to any particular instantiation (any particular neuron in your brain).

In the last half century a great deal of work has been done to trace but also critique the epistemological processes of science. I want to reposition this work in terms of how science brings into view these different scales both through intervention and explanation. In the tradition of delineating metascientific models of science, we can generalize the accounting or marking of scale by considering all science in terms of specification.[12] Here we mean specifications both in the sense of making something precise and in the technical sense of detailing the particulars of a project or technical object, including where and how it might function. The goal here is to see if we articulate more precisely both the valuable function and the limits of science's mode of inquiry.

To make this concrete, let's pose another simple example: here is a red bell pepper. Explain it! With an object assumed, we can proceed via descriptions attached to the object. While we can explore the pepper's this-scale qualities to attach to it things like taste, color, uses in cooking, which of my children like it, and so on, we can quickly find ourselves with scalar specifications. Where did it come from? We find ourselves describing the grocery store, and this pulls us into an economic system described by price, distribution systems, modes of food regulation, and so forth. We can trace it to the farm, and quickly move into large-scale or small-scale ecological relations, the thermodynamic and chemical needs for it to grow, the qualities of the soil, the climate conditions. Remaining with the pepper, we can go to molecular attributes to identify why it's red or has the texture that it has, explain these using evolutionary pressures such as long-term animal-pepper relations—and so on. We can imagine each of these as a parameter attached to the pepper in a kind of specification chart containing the many different ways to explain this pepper. These specifications say a lot about the pepper; they help us identify it, trace its relation to other objects, explain its characteristics according to particular parameters, and tell us ways it might be used.[13] In this form, all of these aspects are presented as facts about an object already assumed; we can therefore call them *ontological specifications.*

But suppose a skeptic arrives at our pepper and asks, "So you say that this pepper is red because it contains something called capsanthin. What is that? How do you know? Does it apply for all peppers?" This kind of question prompts us to augment this chart of specifications-as-facts with specifications about the knowledge process itself, which unfolds into a new set of delimitations about these facts and explains their genesis, justification, relation to each other, reasons for definition, and limits of applicability. These *epistemic specifications* are *about* the first sort of specifications; they are about the process of coming to know. This distinction between the ontological and epistemic specifications becomes embodied in the distinction between the textbook or reference book (such as the Merck Index for chemicals) and the messy practices of science. However, it is in working through epistemic specifications that science is practiced. But this creates a problem: the information about the acceptability of a description, how it was formed, and to what it may apply are all contained in the epistemic specifications. As we saw in a more general sense in chapter 7, ontological presence assumes too much already about the conditions under which that object is formed and described. Both what we call facts and standard knowledge then hide, in a sense, the epistemic specifications within these statements about the object. Doing so can be useful; the Merck Index, a protein database, and a taxonomic classification all provide ways for scientists to build on previous work without having to repeat that work (see 3.11).[14] Perhaps this does not matter for something like capsanthin in a red pepper, but the tension between the two kinds of specification arises repeatedly at points of controversy, political intervention, when some particular scientific object catches the attention of nonscientists (such as monosodium glutamate), and, as our fourth description indicates, when we attempt to bring that statement back to the pepper in front of us.

The situation becomes more marked when the object in question is from another scale, as capsanthin already is, since the object itself is not brought before us except as a series of specifications gathered around this term, some of which are ontological, some epistemological. Such "objects" immediately invite wonder: how did we find these things out? What do they apply to? What do they have to do with these things here? Scientific practice, it might be said, is the process of developing, sorting out, and making use of these specifications in a way that aims to account for and regulate these epistemological processes so that scientists can say with some precision how and in what context such descriptions apply.

However, out of this epistemological process emerge ontological specifications, that is, facts about objects. Given the power and ubiquity of scientific knowledge, it only makes sense that there would emerge scholars whose goal is to trace out these epistemological processes so as to make clear the process of making these facts. If scientists expect people to change their behavior and worldview in relation to things like DNA, buckminsterfullerenes, climate change, viral vectors, or our struggling microbiomes, then these specifications are not simply a matter for scientists. This is not simply a matter of acceptability, authority, or truth: it's about making these objects acceptably real, so that we can trace more closely the ways that such specifications may or may not become the basis for new relations, new forms of speech, and new worldviews. This is how I would view the emergence of the various fields grouped together here as science studies.

As science studies has matured, it has provided a rather extensive view of the scientific process. In surveying this literature, we can define the following modes of epistemic specification.

1. The object/epistemic thing—what we defined above as a given object exists in science primarily as an epistemic thing that emerges in and becomes subject to an experimental system.[15] Such objects are not necessarily assumed in advance, but emerge as a science develops. Scale clarifies that this specification of the epistemic thing must occur first, since scalar objects cannot be taken as obviously present.
2. The apparatus—the device(s) used for tracing differences that make a difference, particularly those on another scale. One must know the specifications of the apparatus in order to be able to use that apparatus as a means of testing hypotheses and descriptions. This includes the accounting for quirks of the device, particularly potential distortions produced by it, and the modes of intervention used to produce knowledge.[16]
3. The paradigm—the dominant theories and domains of acceptability that guide both questions and interpretations for an inquiry. Thomas Kuhn's original description of paradigms remains a powerful way of considering how scientists delimit what

questions are relevant, how to interpret data, and what kinds of descriptions are acceptable.

4. Discursive productions—everything from scribbled notes and back-of-the-envelope calculations to conversations in the lab, professional meetings, academic publications, and popular accounts. One can specify how particular aspects of these descriptions relate to, frame, alter, guide, and otherwise create aspects of a science.
5. The scientific culture, or social aspects of scientists—their day-to-day activity, modes of interpreting evidence and rallying behind interpretations of that evidence, trainings, investments, personal biases and quirks, and so on.
6. Relevance for human action—including possibilities for technological development, questions of research support, habitual biases and omissions arising from the cultural milieu, and political exigencies, interventions, or restrictions. One can specify, for example, how the funding priorities of a national science funding agency or special interest group affect what is researched, how it is presented, or even how it is practiced.
7. Experience—as in the fourth description of a cell. One can specify how scientific inquiries relate to the possibilities of experience.
8. Existence—here I want to separate all of the others from the ground of being in order to acknowledge that science is a process of specification out of what is. How is "what is" being resolved? Separating existence from both object (epistemic thing) and apparatus resituates the question about the reality of the things science describes: the cell as an epistemic thing is already the result of a specification. Thus, by "existence" here I mean something abstract only in the sense that it is abstracted from any one object and observer. *The specifications in this category navigate the relationship between objects and observers.* Thus, this is where I would place scale, as one essential specification of how reality is being examined, described, or understood. Scale is not quite a specification about the apparatus, since it is tying the differences discerned in the apparatus according to their relation to existence apart from any given object. Likewise, scale must be ambivalent about objects themselves even as it traces and relates epistemic things. Scale's simple abstraction away from any given

object or situation likewise means that it also cannot be a specification about discursive productions, human action, or culture. Finally, even as it is tied to experience, its power lies in the consistency whereby it could transform experience by relating it to knowledge gained outside of immediate experience. Thus, scalar specifications function as parameters for the others, relating together epistemic things, apparatuses, human-bound interests, and experience to a reference point about how whatever "is" is being resolved.

These modes of specification emerge more or less simultaneously in science. From the perspective of scientific practice, these are modes of becoming more precise and ensuring that descriptions are adequately understood and applicable. The problem is that, in experience, one already has an apparatus, that is, the senses, which parses reality in a particular way at a particular scale. Science expands these capacities outside of the domain of any single, unaided *Homo sapiens* body. In doing so, practices and devices must be developed for clearly tracing and regulating these extensions. This requires diminishing, regulating, or accounting for the role of individual experience, the culture of science, relevance for human action, and description (specifications 4 through 6) so that science can account for how the apparatus adequately specifies the object within the paradigm (specifications 1 through 3). This operation is necessary *because* these objects of inquiry extend beyond human experience. If descriptions were not regulated by specifications in the apparatus, object, and paradigm; if inquiries were not insulated, to some degree, from human interests; and if science did not abstract away from any particular experience, then we could not trace out these inquiries arriving on another scale. How, then, could we speak with any confidence at all of cells or climate change or DNA or viral transmissions?

THE DISCIPLINING OF SCIENCE (STUDIES)

The products of science—whether textbooks, colliders, television shows, DNA evidence, or political recommendations—arise from the tracing of these modes of specification. In the process, science slows us down to examine our claims and to provide careful parameters for our descriptions. This is why the philosophers Gilles Deleuze and Félix Guattari describe science as "a fantastic slowing down" that defines "functions presented as propositions in a discursive systems"—what they call "functives."[17] Science may, as Deleuze and Guattari characterize it, "provide chaos with reference points," but only to the extent that it traces and retains these modes of specification.[18] Here we catch wind of the critique implicit in much of science studies: one concern about this mode of specification is the way it inevitably excludes and precludes. The danger is assuming that these modes of specification *contain everything we need to know about reality*. This is the problem with converting these epistemic specifications into ontological statements: conditions of knowledge and domains of applicability become buried in statements of fact.

Given this danger, we can understand a certain attitude arising in response to the excessive authority and weight given to science. A distressing characterization can be found, for example, in Martin Heidegger's argument that "science sets upon the real," making existence "stand in objectness" so that "at any given time the real will exhibit itself as an interacting network, i.e. a surveyable series of related causes."[19] Heidegger relates this practice to his critique of technology, arguing that science performs an "enframing" that results in a "most extreme dominance" in which the world "becomes a standing-reserve to be commanded and set in order."[20] Even in this critique, Heidegger largely buys into the presumption that science's mode of specification captures and totalizes Being. We can step back from this distressing characterization by speaking of science as a specification process that does not capture relations but rather maps and traces them for particular ends in particular, limited contexts. In fact, *these limits—that is, its own specifications—are the very reason science functions with any authority at all*. This may be what decades of science studies work has made possible: a renewed account of the specification processes of science, not to undermine that process but to further specify the domains of applicability of scientific knowledge.[21] This is what we did in the fourth description of a cell: point to another configuration of description that brings us outside of the typical scientific specifications but nonetheless bears a relationship to those descriptions.

Given the suspicion in Heidegger's characterization, we can see how this tracing of epistemic specifications can be situated as an attack on the authority of science. The resulting conversation has hidden somewhat the aspects of science that I want to highlight here, particularly an account of the nature of the not-knowing embedded within science. Some of the difficulty is an artifact of the response to the growth of the authority and scope of science, as well as the way science was reified as a special form of knowledge. Both the sociologist David Bloor, who was instrumental in the development of the sociology of scientific knowledge, and Bruno Latour, for example, were working against the way that sociologists avoided studying the foundations of scientific knowledge by making science a special case, insulated from cultural productions and different from other kinds of knowledge.[22] Similarly, the strongest studies on political aspects and implications of science arose from feminist critiques, which then developed into particular interventions into the ways that political interests or social bias affect science.[23] These studies are responding to a conception of science as objective, with exceptional or sole access to truth void of social, personal, or political interests—a view that gave science priority in both funding and interest and emphasized scientific expertise over and against other forms of expertise. The resulting tenor in such work is inherently critical, positioned as a task of demystifying, unmasking, or unveiling the practices and claims of science.[24]

Latour and others have argued that this critical tenor undermines the hopes that science studies would actually strengthen and bolster science. However, the basic operation of criticism is essential for tracing out and disentangling these epistemic specifications. The problem is that, when situated as an epistemological critique demystifying and complicating the authority of science, science studies scholars such as Latour and Bloor often overstated or confused the nature of this retracing. To clarify, reposition, and defend critical methods, I want to use what we have said about scale to pinpoint three potential confusions that arise when we examine the limits and operations of science. We can take reference to Latour's widely cited philosophy, method, and characterization of science studies to present one version of these problems.

The first confusion is the reprioritizing, inverting, or muddling of the kinds of specification. In a critical mode, this maneuver may take the form of "we think scientific knowledge is unaffected by X but actually it is." In this form, the scholar takes an aspect of the epistemic process of science—social relations, cultural forms, rhetorical productions, political influences, ideological biases—and shows how they are necessary for the production of facts. Doing so is not a problem in itself; the problem is when the claim reprioritizes these specifications, suggesting that the social *creates* scientific knowledge. This maneuver was essential, for instance, in Bloor's work on the sociology of scientific knowledge; his argument was that, because all knowledge is created through human social institutions, one should position social practices as a, or even *the*, causal factor in how scientific knowledge develops.[25] The cause of Pasteur's germ theory of disease is not just the microbes he is observing but the social practices through which his theory is articulated.[26] When situated too strongly, this becomes a crossing of domains of specification, counterposing the social specifications under which a fact is accepted with the process of tracing the differentiations on which those social deliberations are based. This problem is implicit in all framing of "social construction": such studies usually attempt to substitute the scale of the social as the cause (or at least a cause) of something.[27]

In a series of exchanges, Latour critiques Bloor and others for just this problem, but he does so through an equally confusing operation: muddling the forms of specification by placing them all in the same terms. In Latour's account, all of the operations of science become "trials of strength" by "actants," which include nonhuman aspects of reality. Thus, while Latour refuses the prioritization of the social, he does so by essentially generalizing the social. The language he uses is social and political through-and-through, leading to a confusing articulation of the processes of science as it piles together all of these different aspects in the same terms.[28] As he develops this maneuver, Latour argues that he is moving from a distinction between "what is constructed" and "what is not constructed" to "what is constructed *well* and what is *badly* constructed."[29] This move to quality of construction, combined with the generalized language of sociality, muddles *the kinds of construction* that we're dealing with. Since all of the epistemic specifications are piled together into this "parliament of things," Latour gives little space for prioritizing the reflective, filtering practices science uses to make clear and discernible things that are not easily traced out (most importantly, scalar objects). I would suggest that his defenders of science who are antagonized by his critiques are attempting, even if they do so poorly, to distinguish between those parts of science that are constructed by the interaction of nonhumans and those that are constructed by the interactions with humans. It only adds to the confusion to say that it's all equally constructed—but, by the way, these constructions also include those from nonhumans. While we undoubtedly need to acknowledge the role of nonhumans in producing science, our scheme of specifications implies that one can distinguish a kind of construction that derives its form primarily from an encounter with an object, with other domains of specification adequately accounted for. Is this not what Latour's naive scientists are trying to say when they say "it is not constructed"—with an implicit "by us"? Rather than characterizing these scientists as foolish, backward moderns, perhaps those of us trained in the nuances of language, rhetoric, and culture can assist our fellow inquirers in clarifying the domain and form of their own operation.

The second confusion is the philosophical maneuver that underlies Latour's generalizing of sociality: as he makes clear in his exchanges with Bloor, Latour is arguing that scientists aren't just creating knowledge, they are creating objects. Thus, while Bloor claims that social institutions are responsible for creating knowledge about microbes, Latour makes the stronger claim that, in some sense, science creates microbes.[30] This is why Latour's maneuver is "one more turn after the social turn": "both [Nature and Society] are the results of the practice of science and technology making."[31] Or, as he says in response to Bloor's bafflement with this statement: "Every single one of the philosophical difficulties of the moderns comes from this double bind: driving a wedge [between subject and object] and then, after having rendered problematic the connection, trying to patch it up."[32] What we have said about scale, particularly in Part II, indicates that the wedge between subject and object needs to be reexamined, but Latour here gives too much power to these very moderns he critiques. Some "moderns" did not create the subject-object distinction. Latour places this split too late in the epistemic situation. *Any experience produces a subject-object distinction* (see chapter 8)*; this is not something uniquely formed in the scientific process.* While Latour does not mean to attack science in this maneuver, in an important sense he is, because accounting for epistemological conditions of science cannot proceed as if *science is the one creating the split between nature and culture, inside and outside.* Systematic forms of inquiry, including science, are designed to examine and trace how and where these divisions are made. We might agree with Latour's essential point that the configuration of objectivity and the real "out there" is misplaced, but it does not clarify the issue to suggest that science constitutes this difference. Instead, science can be seen as an attempt to account for and systematically produce particular subject-object configurations through these modes of specification. In this view, science is primarily a reflective endeavor: it experiments with its own descriptions and apparatuses in order to see how and where they function. It then uses these specifications to rally parts of the world (humans or otherwise) into particular configurations. While this language may drift close to Latour's actor-network language, the view is equally close to Bloor's, who argues that the subject-object distinction is necessary to account for how any interaction is configured.[33]

The third confusion arises in how Latour remains excessively committed to this question about truth and validity of facts, which, I would argue, obscures underlying questions about the ambiguities and limits of scientific description.[34] Of course, we need to trace science's modes of specification. But when Latour moves to the broader questions about the formation of the subject-object distinction, the acknowledgment of nonhuman agency, and the construction of reality, he pushes science studies into territory beyond truth conditions while still insisting this is about the epistemic process of science. Situating this analysis in relation to truth conditions takes this "modern constitution" too seriously, as if it is only in the modern era that such diagrams of nature and society are possible, as if they are generally accepted, rehearsed, and understood, and as if there are not already alternative ways of conceptualizing such knowledge practices. Doing so places science studies too squarely on the same ground as science: both being concerned primarily with the validity of the specifications produced by science. There are, however, far more issues that arise from these specifications. Most apparently, there is the issue of the limits and applicability of these very specifications. In addition to questions about the particular conditions out of which science arises there are also questions of how it makes possible new perspectives and whether and how science results in further transformation of worldview for those engaged with it. In response, we can acknowledge the inherent limits and ambiguity of the very process of specification. In doing so, we can move from questions about truth to other kinds of questions about what these discourses do in different contexts. Is it not different to ask "What is a cell?" in the way we have here? Perhaps this may seem a strange move in a time when truth, accuracy, and science are under assault, but I wonder how it would change the conversation about science if we began with a different question: "What are we talking about?"

FROM SUSPICION/SKEPTICISM TO BEWILDERMENT/WONDER

This last option is not readily apparent in Latour's philosophy, even as he moves from "critique to concern" in his reckoning with the science wars. This simple question—what are we talking about?—points us to the ways that scientific descriptions pull us out of the scale at which we live and require that we adjust our *this-scale* behavior on the basis of this reconfiguring of reality. On the one hand, this is all the more reason we need to figure out new ways to keep intact these matters of fact, so that we can make these extensions with some degree of confidence. On the other hand, these matters of fact are not enough. They become matters of concern not because they are constructed equally by other actants but because they run into tension with a this-scale, human way of understanding reality—when I have to reconcile the struggles of my life (paying rent, keeping a job, raising my kids) with the confusing scalar implications that connect together one scalar object (economics) with another (ecology) in a way that is not immediately apparent and easy to grasp.

In this light, we can wonder about Latour's move to concern. While I am sympathetic with his echoing of Haraway's aim for a discourse designed "to protect and to care," "matters of concern" seems to also imply an overbearing scrutiny regardless of whether or not the thing is clearly understood—or even if we're certain that we're talking about the same thing.[35] Is this not already what happens when science becomes entangled in a concern—about vaccines, climate change, women's bodies, health-care systems, fracking sites, clean water, or mask mandates? Latour argues that facts are closed while points of concern are open, ready to be composed and recomposed. Yet both facts and these points of concern run into the problem of *doxa*, the marriage of opinion with social expectation, where everyone proceeds as if the issues at hand are clear and the points of concern are shared. In addition, we can follow Latour's concern about critique and consider the affect here: does concern not imply worry, of being concerned about your neighbor, certain that something is wrong, that I must get involved and intervene? Is this not the unfortunate result of this insistent sociality? Like parents concerned about their teenager who has been staying out late—and have you seen what he's been wearing? (and he spends so much time alone)—we find ourselves plagued with an excess of concern as we contemplate our scalar entanglements. Is this the attitude with which we will approach our dear Gaia or our fellow actants?

In this light, we can further reposition the idea of critique and push against Latour's critique of critique in order to rediscover this more strictly critical maneuver: discovering what is not understood in what we think we understand. Latour's characterization of critique is both egoic and suspicious, as if the critical move is inherently (1) about me, the critic, getting the upper hand, being more informed, and less naive, and (2) wary or even disdainful of the thing that is being critiqued. Certainly, there is a real target here (Latour puts his early work among them), and we need to scrub criticism of this attitude.[36] But Latour uses that characterization to throw out what remains an essential move: attending to the conditions, parameters, limits, and implications of a mode of knowledge. Critique does not have to be suspicious; it does not need to reduce an object to something else (e.g., your distress to a death drive); it does not need to be an ego trip where you are the enlightened one bearing the truth to the ignorant. Latour mixes up the general tenor of "being critical" with the Kantian operation he insists that we get beyond: the task of observing how, within any particular experience, there are conditions, assumptions, and components that we may not be aware of but that can and should be examined.

It simply is not true that "what performs a critique cannot also compose."[37] Criticism is needed because *we already suppose too much about reality*. There are already social practices, languages, cultural regulations, personal sentiments, technological infrastructures, genetic inheritances, energy grids, oil sources, residual wastes slowly dissolving in the ocean—all of which must be examined and reopened for us to forge new forms of existence. Is it not strange that Latour would say that "With a hammer . . . you can do a lot of things . . . but you cannot repair, take care, assemble, reassemble, stitch together"?[38] Need I point out that people do all kinds of repairs and assembling with a hammer? Surely Latour, who would so readily have us rethink our conception of science, politics, and things, understands that if you're going to repave a road, you must first dig it up?

How bewildering! Do we know what we are haggling about? This, it seems to me, is the fundamental question brought on by a basic critical impulse: what is it that we are talking about? Is it as clear as it seems to be? I'm not taking the strategy shared by Bloor and Sokal and simply accusing Latour of obscurantism; rather, I am pointing to a set of more basic questions prompted by a different affect meant to set stage for a different conversation. What is Latour referring to, for instance, by "nature" when he insists that "ecology seals the end of nature"?[39] When he says "all the various notions of nature" does he mean Emerson's nature *and* the one that evolved in frontier America, and the national parks movement, and the one assumed by physicists, and the one environmentalists are trying to save? No. In Latour's own work he means particular notions and stances tied to this term "nature," such as the world conceived as a place apart or an object to be dominated by a knowing subject. Such aspects, uncovered by critique, do not themselves totalize "nature" in a way that warrants such strong dismissals.

In their deconstructive form, these questions arise from bewilderment, in the not-understanding and confusion that is taken as neither mine (my inability to understand, my not being properly trained in your discipline) nor yours (your obscurantism) but as a way past pretending to understand or that there is only one way that descriptions work, experiences are configured, or actions are performed. What are we even talking about? To what does it apply? How might that be understood? How do we know it? What is not accounted for in this articulation or mode of knowing? How can we work with it? In their most productive forms, these questions hinge on wonder: I wonder what X is? What if we try X? How does X relate to Y? Can there be something else here? Bewilderment is the feeling one gets in a tangle, the "confusion arising from losing one's way"; it is the acknowledgment of the labyrinthine form of our attempt to understand where we are, what we are doing, and what we are saying (*OED*). Wonder arises from the astonishment with objects, things, ideas. The two affects are complementary, both moving us to appreciate anew and ask again, more carefully, with less certainty and more interest.[40]

The same questions could be asked of skeptical scientists such as Sokal when they adopt the same strong modes of rhetoric, which assume too much common ground in the understanding of the terms and acceptable forms of proof. Skepticism is the counterpart to critical suspicion; it risks hiding the need to specify, trace conditions, and withhold certainty behind an egoic, superior affect. In this form, as Latour fairly notes, skepticism has been excellent for "debunking quite a lot of beliefs, powers, and illusions," but it is also reaching its limits.[41] The problem arises when scientists make skepticism into an attitude based in authority and mistake their highly regimented mode of specification for the only mode of parsing the world.[42]

By way of illustration, one can wonder what would have happened if Carl Sagan had taken a different approach in the conversation he recounts at the beginning of *Demon-Haunted World*. There, he finds himself in the car with a driver who, upon finding out that he's "that scientist guy," begins to question him about "frozen extraterrestrials languishing in an Air Force base near San Antonio, 'channeling' . . . , crystals, the prophecies of Nostradamus, astrology, the shroud of Turin." With each new topic, says Sagan, "I had to disappoint him. 'The evidence is crummy,' I kept saying. 'There's a much simpler explanation.'"[43] Instead of being skeptical, assuming the role as the arbiter of truth, and simply recounting the evidence for and against such things, Sagan could have asked something different: to what is he referring? I've wondered about channeling . . . what is it describing? Why are we so certain these are spirits? What do we mean by that? Undoubtedly these questions can be disarming, and they probably would have confused the driver. But, if done with more nuance and less ego, this approach can respect that there is some configuration toward reality being expressed here. What is it? These questions, if adequately shared, can induce a mutual bewilderment from which we can start to have a conversation about these different kinds of questions, what they apply to, why we hold to them, and what they serve to do.

It certainly is harder to have this kind of conversation with ideas like Atlantis. But many of the things described by science are closer to Atlantis than they are to the immediacies of one's life—and harder to wrap one's head around. In science, bewilderment is thus the first step toward working with these broader difficulties and reorientations. This is the method of the current project: it began with a bewilderment (how is it possible that I am simultaneously atoms, cells, quarks, ecologies, galaxies?) and a wondering (how does this apparatus function? Why does it show up in everything from descriptions of mystical union with God to descriptions of ecological entrainment? But also: have you seen the "Inner Life of the Cell"?!).[44] The method then proceeded by criticism—a tracing of the conditions under which scale can be articulated—but also through other methods of textual exegesis, historical questioning, comparative analysis, and thought experiment. In many ways, I can hardly get past the first detail, the basic structure of scale; I keep unfolding more considerations out of this fractal question that might simply be reduced to a word, with an affect: scale?!

Yet as the breadth of this inquiry implies, such bewilderment can be productive of new avenues of discussion as we unfold a simple point in the many contexts in which it inevitably applies. It is an invitation to step back from the things assumed (the critical maneuver, but here performed mutually, starting first with the critic's confusion and willingness to admit an impasse) but also to specify (Say more about what you mean by that? I'm not sure that we're talking about the same thing.

I've often wondered . . .). Perhaps such an affect and intellectual stance are difficult to maintain. But I wonder if this is only because we get so involved in the matter, concerned about its implications, its problematic uses, its terrible associations, its naïveté. My experience is that such conversations end in discord primarily because someone, at some point, invokes authority, insists on his understanding, or simply runs out of energy (pick it up again tomorrow?). Is this not a practice for humanists to take up: helping ourselves and others step back from this insistent sense of knowing, to trace these disjunctures, and help ourselves navigate these bewildering transformations of our existence?[45]

If we combine this bewilderment with the four descriptions of a cell, we can notice just how much more there is to ask about what science does and describes. Okay, let's accept a basic fact like cells. What unfolds out of that? To what does each form of description apply? In this rendering, one notices that, as we add more specifications of the object in the second and third descriptions, we not only learn more about the object but also delimit the domain of applicability by embedding these specifications out of which the diagrams are produced. Science studies shows that these conditions can be made available if we ask the right questions. Since science proceeds by adding onto their diagrams more often than deconstructing them, we might think that they are not interested in asking these questions. To the contrary, I want to show that this simply moves the wondering forward.

ON HUMBOLDT'S BRIDGE: THE SUBLIME SCIENCE OF THE COSMOS

Science encodes a not-knowing in a knowing. We can now explain this sentence more precisely: in defining, examining, and describing an object, imbuing this description with particular aspects that operate according to particular specifications (including a particular scale), science creates knowledge. But this positive articulation is delimited; in reality, it functions as an adequate description because it is limited—we know to what it will apply and why. We know that all cells will have some kind of DNA, that some kinds of proteins will unfold this DNA, and we know how, in order for an aerobic cell to generate the energy that it does, it has to follow a certain chain of exchange among molecules. We do not need to know when and precisely how each and every cell has or does these things, because the description operates according to delimited categories such as "cell," "respiration," "mitochondria," and

"DNA." If we were to talk about prokaryotes many of these descriptions would not apply. They would again apply quite differently—or not at all—if we were talking about the carbon cycle around the Susquehanna Valley or the respiration in a fungus.

This point is not new, but scale brings it into sharper focus. To cite a precedent: this is already within Polanyi's notion of "tacit knowledge." Polanyi argues that, simultaneously, "(1) Tacit knowing of a coherent entity relies on our awareness of the particulars of the entity for attending to it; and (2) if we switch our attention to particulars, this function of the particulars is canceled and we lose sight of the entity to which we had attended."[46] Polanyi's ontological version of this is more directly scalar: "(1) that the principles controlling a comprehensive entity would be found to rely for their operations on laws governing the particulars of the entity in themselves, and (2) that at the same time the laws governing the particulars in themselves would never account for the organization of the higher entity which they form."[47] Polanyi is here paralleling the same argument about systems that we saw from Bertalanffy in chapter 9 with an epistemological argument: that the diverse specifications we attach to an object do not capture the whole of it.

To appreciate what this means for science, we can rediscover a notion of science that acknowledged the form and effect of this gap in knowledge and, rather than shy away from it, built science around it: Alexander von Humboldt's "science of the Cosmos." As Laura Dassow Walls has made clear, Humboldt was instrumental in developing nineteenth-century science, despite being largely forgotten in the United States.[48] His groundbreaking multivolume *Cosmos*, published between 1845 and 1862, provides a significant precedent to the scalar terminology and concepts used here; my use of the term "Cosmos" reflects his usage as much as Sagan's. Walls argues that Humboldt created a way of speaking of "a planetary interactive, causal network operating across multiple scale levels, temporal and spatial, individual to social to natural, scientific, to aesthetic to spiritual" (11). Humboldt's *Cosmos* clarifies the value and method of moving from the aggregate of disconnected scientific facts to a sense of connection that spans multiple scales. His methodology aims to "close the gap between mind and nature," to include everything and see it all together, and a justification of how and why one might do so—all situated in response to the widespread Enlightenment praise of rationality over and against speculation or superstition (9). Thus, we can follow Walls in examining "Humboldt's bridge" between the sciences and other domains of knowledge at the moment before specialization fragmented this approach (10–11).

Cosmos was Humboldt's last work, the culmination of a long career built on a growing network of knowledge, but set to an ambition he had as a young man: "The principle impulse by which I was directed, was the earnest endeavor to comprehend the phenomenon of physical objects in their general connection, and to represent nature as one great whole, moved and animated by internal forces."[49] Humboldt describes the experience of looking out over oceans and flat plains, seeing the vast expanses of things, and feeling that these things are interconnected—even as he participates in the practice of examining each of these separate objects. Contrary to the tediousness of careful scientific observation, the patterns he finds induce wonder: "The powerful effect exercised by nature springs . . . from the connection and unity of the impressions and emotions produced; and we can only trace their different sources by analyzing the individuality of objects, and the diversity of forces" (6). He thus connects together this detailed analyzing of objects and forces with the wonder of a broader, scalar view.

The result of this confluence is that, although we usually speak today of wholeness and unity as contrary to rationality and experience, for Humboldt rational empiricism itself points to such unity: "Nature considered rationally, that is to say, submitted to the process of thought, is a unity in diversity of phenomena; a harmony, blending together all created things, however dissimilar in form and attributes; one great whole (*to pan*) animated by the breath of life" (23). Humboldt's parenthetical insertion of the Greek *to pan* is the essential scalar maneuver we have followed above (1.23, 3.38, and chapter 7). He repeats this insertion throughout *Cosmos*, to remind us that by "the whole" here he means the aggregate nature of the All (51, 62). Include everything, and wonder and interconnection will naturally appear.

Humboldt positions this maneuver in between disdain for poetic sentiments by certain scientists and concerns about the coldness of scientific description by some romantic thinkers. From one perspective, the emphasis on physical observation, experimentation, and mathematical calculation, which he labels "rational empiricism" (30), creates a disparate set of facts that avoids any larger statements about the structure of the Cosmos as a whole: "Physical philosophy . . . when based upon science, doubts because it seeks to investigate, distinguishes between that which is certain and that which is merely probable, and strives incessantly to perfect theory by extending the circle of observation" (18). Rational empiricism relies on skepticism because it aims to investigate what is within any observation. However, as one observes more and more, this empiricism also begins to observe discrepancies to any law trying to describe this assemblage of things. The result is that "instead of seeking to discover the mean or medium point, around which oscillate, in apparent independence of forces, all the phenomena of the external world, this system delights in multiplying exceptions to the law" (18). This is what we have described as the adding on of specifications—the continually adding conditions and particularities attached to any fact.

"Enjoyment" of science, Humboldt argues, arises when one takes into account both the factual correctness and the larger intuitive encounter with this diversity of existence: "The higher enjoyments yielded by the study of nature depend upon the correctness and the depth of our views, and upon the extent of the subjects that may be comprehended in a single glance" (18). This maneuver is based on a scalar shift, using and attending to the details of observation to support and bolster a description, but still keeping in view the larger pattern involved. Thus he speaks of his project as an attempt to "prove how, without detriment to the stability of special studies, we may generalize our ideas by concentrating them in one common focus, and thus arrive at a point of view from which all the organisms and forces of nature may be seen as one living active, whole, animated by one sole impulse" (36). This "common focus" is produced by considering all of these things together at the scale of *to pan*, the All. At this point Humboldt turns to poetry: "'Nature,' as Schelling remarks in his poetic discourse on art, 'is not an inert mass; and to him who can comprehend her vast sublimity, she reveals herself as the creative force of the universe—before all time, eternal, ever active, she calls to life all things, whether perishable or imperishable'" (36). In invoking Schelling's critique of a mechanistic view, Humboldt is affirming how a rationally grounded, empirical understanding of the world can itself arrive at a sublime view of the universe.

Responding now to an antiscience version of romanticism, Humboldt argues that this wonder is not couched in the inaccessible. Rather, he relies on the fact that scale also makes, in some way, the inaccessible accessible (see 2.8). Thus, he disputes Edmund Burke's definition of the sublime as a feeling born from the ignorance of nature (19). By Humboldt's account, Burke's definition led to a fear among nonscientists that the knowledge of science is directly antithetical to the wonder, beauty, and appreciation of nature. In response, Humboldt argues:

> Whilst the illusion of the senses would make the stars stationary in the vault of heaven, astronomy by her aspiring labours has assigned indefinite bounds to space; and if she has set limits to the great nebula to which our solar system belongs, it has only been to show us in those remote regions of space, which appear to expand in proportion to the increase in our optic powers, islet on islet of scattered nebulae. The feeling of the sublime, so far as it arises from a contemplation of the distance of the stars, of their greatness and physical extent, reflects itself in the feeling of the infinite. . . . The solemn and imposing impressions excited by this sentiment, are owing to the combination of which we have spoken, and to the analogous character of the enjoyment and emotions awakened in us, whether we float on the surface of the great deep, stand on some lonely mountain summit . . . or by the aid of powerful optical instruments scan the regions of space, and see the remote nebulous mass resolve itself into worlds of stars. (19–20)

Humboldt replaces ignorance with vastness as the source of the sublime. Since science allows us to discern how things are actually far more complex, vast, and layered than our usual experience leads us to believe, it creates sublimity through its endless increasing of objects included in this All. The reason for this "imposing impression" is this combining force of scale, of taking together all of these details and adding them to the same set (All) until we see a larger sense of unity emerge (3.12, 3.38). Thus, Humboldt adds the telescope to the classic experiences of vastness: the ocean and the mountain.

This new perspective is produced by accumulating descriptions with as much detail as possible without losing this larger perspective. In this way, *Cosmos* is an exemplary accumulation of scientific knowledge. Humboldt begins with astronomical phenomena and ends with descriptions of terrestrial organisms, including some brief descriptions of their microscopic features. It is thus a careful gathering of scientific facts honed to direct our attention to seemingly disparate aspects of the universe, on multiple scales, to demonstrate the overarching unity and interconnection that exists within this vast range of things. But in this multitude of things we must not lose sight of the larger unity: "The principle of unity is lost sight of, and the guiding clue is rent asunder whenever any specific and peculiar kind of action manifests itself amid the active forces of nature" (56–57).

The term *cosmos* captures this practice of description, and Humboldt frequently refers to his work as "the science of the Cosmos" (e.g., 36). Humboldt notes that he is using the word in a new way that unites the Greek use of the term with this mode of scientific description. In this use, he is invoking this order in relation to the All (*to pan*) that might be discerned out of the aggregation of scientific description:

> It is by a separation and classification of phenomena, by an intuitive insight into the play of obscure forces, and by animated expressions, in which the perceptible spectacle is reflected with vivid truthfulness, that we may hope to comprehend and describe the universal all (*to pan*) in a manner worthy of the dignity of the word Cosmos in its signification of universe, order of the world, and adornment of this universal order. May the immeasurable diversity of phenomena which crowd into the picture of nature in no way detract from that harmonious impression of rest and unity, which is the ultimate object of every literary or purely artistical composition. (62)

The prayer-like shape of this invocation is a plea for us to rediscover the sense of wonder at the wealth contained within the Cosmos. The first sentence here places Humboldt's work within the realm of science, while the last places it within the realm of poetics, positioning these two together in relation to the variety of facts provided by science.

RHETORIC, MYSTICISM, AND OTHER TRADITIONS OF TRANSFORMATION

While Humboldt's "science of the Cosmos" converts science into wonder, I am not suggesting that this is the primary task of science or scientists. Do we have a name for this kind of inquiry, which dwells with such bewilderment, considers its form and contours, wonders at and about its transformation of our experience and capabilities, and develops these into a new appreciation and new stance toward the Cosmos? What kind of inquiry is built on the not-knowing required to observe this situation in the broadest view we can hazard and to sit with the disorientation implied?

Here we come to two main targets for scientific skepticism: rhetoric and mysticism. Perhaps we could add a third term, "philosophy" in its most nonlogical, transformative sense, if we can locate a mode of philosophy geared toward effects of articulations rather than logical or conceptual validity, toward *gnosis* rather than *epistēmē*.[50] Regardless, we can locate all three terms in Plato, who lays out these sets of questions and problems in his texts on rhetoric. Reading Plato's concerns about rhetoric through a contemplative concern for the possibilities of transformation provides a way of conceptualizing these traditions in relation to the kind of approach I've been taking here.[51]

But first, let's return to the 1990s and note that mysticism was a major source of concern for scientists wary of a growing pseudoscience and antiscience. Sagan's chapter on antiscience in *A Demon Haunted World* begins with an epigraph from two New Age writers that combines the language of mysticism with postmodernism: "There's no such thing as objective truth. We make our own truth. . . . If an idea feels right to you, it is right. . . . Science itself is irrational and mystical. It's just another faith or belief system or myth, with no more justification than any other."[52] This painful pastiche combines humanist notions of agency with the languages of postmodernism and mysticism in order to undermine science. This sentiment, combined with the list of supernatural phenomena Sagan cited above, provides grounds for attacking mysticism. Yet even though Sagan acknowledges that mysticism may, like quantum physics, require extensive training to understand, he does not ask the basic question: what is mysticism exactly? Instead, he sidesteps to the scientific testing of a shaman's claim to cure people.[53] If, as we discussed in the Introduction, mystics are those who experience a transcendence of subject-object in a transformative scalar experience, then this has as much to do with healing as race has to do with intelligence.

A sketch of the other target: in 1990, Alan Gross published the widely reviewed and widely criticized *The Rhetoric of Science*, which became a poster child for overinflated claims and interdisciplinary overreach. Gross was particularly critiqued for his argument that science is rhetoric "without remainder."[54] This claim combines the basic premises of the linguistic turn in philosophy with some of the strongest claims from science studies, but tied to the term "rhetoric." While this maneuver may seem strange to those only familiar with rhetoric as a negative term, Gross's claim coincided with a rethinking of rhetoric in the academy in the United States, where it had been both banished and preserved in the teaching of writing (in English departments) and speech (in communications departments). The central maneuver in this rethinking—that truth might be constructed in modes of communication—is not foreign to science studies; the problem was largely this term "rhetoric." Unfortunately, Gross inserts it as a master term, even at one point stating that the work of Latour and Stephen Shapin "is rhetorical criticism in all but name."[55] The result is resistance on all fronts, from philosophers of science who trivialized Gross's notion of rhetoric, sociologists who resisted this renaming of their discipline, and scientists who saw the book as another attack on science.[56]

And from other rhetoricians. In 1993 Dilip Gaonkar published "The Idea of Rhetoric in the Rhetoric of Science," which resisted this expansion of rhetoric.[57] Gaonkar traces this expansion in detail, noting that rhetoricians have gradually made rhetoric an interpretive rather than practical activity (27) but, in doing so, had left themselves with a critical language that was so "thin and abstract that it is virtually invulnerable to falsification" (33). Perhaps most damning is the way that Gaonkar positions the whole maneuver as merely an attempt for rhetoric to colonize and dominate all forms of knowledge (26–27, 34–36). The question here is both about the term "rhetoric" and the methods that might be gathered around that term given its history. Gross was chastened by this critique, editing a collection of responses and substantially revising *The Rhetoric of Science* in 2006 to rework his most controversial claims.[58] In the meantime, rhetoric of science continued to develop from, to use Gaonkar's own terms, an "uncoordinated research initiative carried out by a handful of committed individuals" to a field much more akin to other disciplines within science studies in that it displays "internal variation in theory and methodology" (40). This subsequent work, as rhetorician Randy Harris notes, proceeded despite Gaonkar's critiques, creating an extensive disciplinary dialogue and engaging in broader science studies conversations.[59]

I recount this conversation about the rhetoric of science to highlight these questions about the notion of rhetoric. Both mysticism and rhetoric share a widely negative association counterpoised with great enthusiasm by those who adopt the terms, both in their academic resurrections in the twentieth century and in the contexts in which they were born. While it may seem an exorbitant task to recover both traditions at once, briefly discussing the two together here will clarify the method central to this inquiry on scale. Most important, mysticism and rhetoric represent two crucial areas in which we easily pretend that we know what we're talking about, even as they function as pervasive straw men both popularly and academically. So again, let's be more precise by returning to what is arguably the source for both rhetoric and much of Western mysticism: Plato.[60]

The standard interpretation, extant already in ancient Greece and Rome and widely repeated by both rhetoricians and their critics alike, is that Plato is *the* enemy of rhetoric.[61] Somewhat unconventionally, I'd argue that Plato's philosophy, when combined with his mysticism, has a positive philosophy of rhetoric that is built on concerns about the power of language. He is specifically concerned about the way we become *intoxicated with language,* which we can read here as an exemplary form of holding-to this-scale experience and values. Plato leaves open a kind of rhetoric (pointed to in the *Phaedrus*) that aims to address or keep in view this intoxication while attending to how language shapes the mind. In the language of this chapter, this amounts to addressing the overinvestment or absorption in your acquired mode of specification. Such concerns are harder to see directly in the standard dialogues read by rhetoricians—the *Gorgias* and the *Phaedrus*—but become clearer in the larger canon. The mysticism expressed in the *Seventh Letter*, the famous cave metaphor in *The Republic*, the *aporia* of knowledge in *Theaetetus*, the chariot metaphor in the *Phaedrus*, and the obsession with "know thyself" throughout—as well as Plato's centrality in mysticism following the Neoplatonists—all make possible a reading of Plato in terms of mysticism that also transforms his view of rhetoric.

In the *Seventh Letter*, the central philosophical passage comes when Plato is recounting a visit to the tyrant Dionysius. Plato wants to see if he is "on fire with philosophy," as he has heard, or whether he was one of those whose "heads are full of half-understood doctrines" (340b). He notes, akin to the discussion of divine madness in the *Phaedrus*, that those who hear of philosophy hold fast to "the daily discipline" and undergo great effort to grasp it (340d), but others "persuade themselves that they have already heard the whole of it and need make no further effort" (341a). "Philosophy" here is not a body of knowledge but a transformation that also requires that one becomes "akin to the object" (344a).[62] The problem, as Plato soon sees in Dionysius, is the tendency to assume knowledge "because of what he had heard others say" and therefore "pretend to a knowledge of the problems with which I am concerned" (341c). Plato then makes what seems to be a strong statement about writing, but which is a standard trope of mysticism: "There is no writing of mine about these matters, nor will there ever be one. For this knowledge is not something that can be put into words, like other sciences; but after long-continued intercourse between teacher and pupil . . . suddenly, like a light flashing forth when a fire is kindled, it is born in the soul and straightway nourishes itself" (341c-d). Writing does not tend to produce such transformations except in those who "with a little guidance [would] discover the truth by themselves" but instead tends to produce "ill-founded and quite unbecoming disdain" in some and "an exaggerated and foolish elation" in others (341e).

Plato goes on to elaborate on this kind of inquiry in relation to names, definitions, images, and knowledge (342b), with a "fifth" topic being the "in itself." In acquiring this fifth, he says, one must be aware of the "weakness of language," particularly in "a form which is unchangeable as is true of written outlines" (343a). In the end, "only when all of these things—names, definitions, and visual and other perceptions—have been rubbed against one another and tested, pupil and teacher asking and answering questions in good will and without envy—only then, when reason and knowledge [*phrónēsis*, a mindfulness, thought] are at the very extremity of human effort, can they illuminate the nature of any object" (344b). If the object here was actually a circle (the example used), then this would be something less than the mysticism that transcends subject-object; but if the object here is that which is at the "extremity of human effort," then it suggests not an object so much as a whole transformation of perspective, much more akin to the glimpse of heaven described in the chariot metaphor in the *Phaedrus* or in the leaving of the cave in the *Republic*. In short, the *Seventh Letter* points to a kind of transformation that requires a different approach to any articulation: you have to test and examine it within oneself (one's experience, thought, world) rather than treat it like an articulation and, to do so, must admit to a not-knowing, dwell with and prod the articulation, and resist the solidity of first impressions, sense variations, and definitions until a kind of transformation occurs that sheds light on the whole situation. This is the attitude and intellectual approach that I have attempted to take here in examining scale.

This framework provides a different way of reading Plato's attack on rhetoric in the *Gorgias*. First, we can resituate Socrates's questions about knowledge (does a rhetorician know how to build a ship? If not, how then can they speak of whether we should build ships?—see 454d). "Rhetoric," says Socrates, "doesn't need to have any knowledge of the state of their subject matters; it only needs to have discovered some mechanisms to produce persuasion, in order to make itself appear to those who don't have knowledge that it knows more than those who actually do have it" (459c). This well-trod distinction between knowledge and ignorance can be read now as a question of transformative understanding (*gnosis*) rather than simply a question about *epistēmē*. After all, Plato is not trying to teach us about ships or any other specific kind of knowledge. The problem is this pretending, this *failure to grasp the limits of one's articulation and knowledge* and proceeding nevertheless. The major issue is that, as he says in the *Theaetetus*, "their very ignorance of their true state fixes them more firmly therein" (176d); they remain like "the blind leading the blind" (209e); he, as Plato adds in the *Sophist*, "thinks he knows the things he has only beliefs (*doxazei*) about" (268a).[63]

The term *doxa* is less central in the *Gorgias*, but it is essential to Plato's concerns about rhetoric. *Doxa* is usually translated as "opinion," but it also indicates popularity, repute, or social appearance.[64] In the *Phaedrus* it appears at the crucial juncture in the chariot metaphor: those who do not glimpse the truth (*aletheia*) "will depend on what they think is nourishment—their own opinions [*doxastē*]" (248b). The problematic rhetorician, in turn, simply studies "what the people believe [*doxas de plēthous*]" (260d), indulging these opinions in order to persuade. This is what is described in the *Gorgias* as "flattery," being "clever at dealing with people" (463a). But the nature of *doxa* is that it requires that you bow to the socially accepted positions and indulge commonly affirmed opinions. This is the core of Socrates's argument about the tyrant with Polus and Callicles in the *Gorgias*: the role of *doxa* is repeatedly performed by the incredulous exclamations of Polus and Callicles, as they defer to the crowd or to popular opinion rather than consider the points of discussion themselves.[65]

However much Plato has Socrates rail against rhetoric in the *Gorgias*, the whole dialogue is obviously about the conditions of possibility for persuasion—of actually having substantive transformation in thoughts, beliefs, and actions. The Gorgias-like rhetorician supposes, through his mechanisms, that he has the power for such transformations, but he only makes use of existing opinions without understanding the situation or actually changing any minds. Most noticeably, *no one is persuaded at all in the* Gorgias *except Socrates*, who states at the end: "For my part, Callicles, I'm convinced by these accounts" (526d). Why? Socrates resists *doxa*, doesn't pretend that he knows, examines the questions at hand for himself, doesn't defer to the common understanding, and dwells with these concerns at length. He doesn't settle for what is "recited in public without understanding and explanation," but looks for "what is truly written in the soul" (*Phaedrus* 278a). This is certainly a problem for science as its descriptions become recited in public, accepted as *doxa,* and invoked as if they should be accepted on authority. Who actually is transformed by these new scales of description? How do we rewrite our souls to accommodate this new situation?

This task arises not from a disdain for rhetoric but from a view of rhetoric as *psukhagōgia,* the leading along of the soul (261a, 271c). This is a different endeavor in both attitude and form from the ignorant rhetoric critiqued in the *Gorgias.* Thus, here at the outset of the concerns about rhetoric we find a description of a rhetoric that is geared toward leading along minds but does not assume knowledge (in the *epistēmē* sense) and respects this power of language, particularly its ability to intoxicate its users into thinking things are understood when no significant reorientation (*metanoia*) has occurred. Indeed, in the context of the chariot metaphor, the whole task of leading along the soul hinges on comprehending the nature of our contingent situation, where we can find *aletheia* (the unfolding of Being) rather than *doxa,* a substantive transformation (*metanoia*) rather than mere affirmation or belief (*pistis*).

In this view, rhetoric would be the tradition that deals not with concepts qua thought but rather with communication as effect, one that assists us in developing awareness of and greater response-ability to these various manifestations of language and symbols. Rhetoric would not merely be about conveying the truth discovered elsewhere—as Aristotle makes it in his *Rhetoric* and as is implicit in the usual distinction between scientific practice and science communication. Rather, it would be about analyzing the ways our articulations work, bringing our attention to this intoxication with language and *doxa,* attending to these possible effects, observing what our descriptions say and don't say, examining how articulations might work differently in different contexts, and similar operations.

This difference is clarified further if we consider how Plato situates rhetoric in relation to dialectic, which he describes in the *Phaedrus* as the cutting up of a topic into its "species along its natural joints" (265e). The dialectician considers and examines divisions so as to be clearer and more precise about the language we're using, while the rhetorician considers these divisions from the perspective of their effects—how they lead along our minds. If philosophy as a domain of inquiry has become primarily dialectic, the development of concepts and the tracings of logic, and science has also become a dialectical process of providing and regulating divisions and descriptions on empirical grounds (what I have here described as specification), these do not encompass or account for this additional task: the question about where and how these descriptions function to change our ways of thinking, speaking, and acting. Likewise, in a more contemplative mode, science and certain kinds of philosophy add to our concepts and descriptions according to particular specifications rather than situate these descriptions within our experience and worldview. So while the rhetorician does not need to pretend to a knowledge of these things (I don't need to pretend to know everything about how a cell works), the rhetorician would both examine and experiment with these articulations as they extend in all kinds of contexts and implications, whether personal, political, cultural, or otherwise. Indeed, there is an ever greater need for this task, since, in the double movement of increased specialization and more widespread use of scientific terms and knowledge, the risk of simple acceptance of articulations on the basis of *doxa* increases.

In this guise, the critique that rhetoric has no subject matter of its own is a strength if the form of expertise is a capacity to attend to the effects of articulation. It is through an awareness of the limits of knowledge that one can attend to such effects. Scientists have to know precisely how their descriptions apply in a constrained context according to an accumulation of specifications. The knowledge problem identified in the *Gorgias* is about failing to see the specifications of knowledge: that cooking does not bear the same specifications—it does not apply to the same things—as medicine, even though it appears to be about the same thing. Rhetoricians who ignore these kinds of problems miss the very task of attending to the power and limits of the descriptions they are using. But these forms of knowledge do not exist solely within those specifications (the first three descriptions of a cell) but must, in order to function, extend beyond them to the places where they are meant to apply (the fourth description). Such work does not need to begin with facts as things needed to be conveyed to the ignorant masses, but rather with facts as specified modes of prodding the world that need to be translated out of the particular contexts of their formation so that we can transform our perspectives according to this new information.

NOT-KNOWING EVERYTHING

Sagan carries the spirit of Humboldt's *Cosmos* into the visual medium of television, where the scientific aggregation is rendered visually into a scalar form. Since we cannot actually see the view of the observable universe or the scale of atoms except in limited forms, Sagan's "ship of the imagination" helps navigate these nonhuman scalar thresholds. While Sagan is operating from the same basis as Humboldt, the visual medium assists but also risks hiding these transformative scalar elements.

Consider one of Sagan's favorite images, in which whole galaxies and groups of galaxies are shown to look something like blurred stars. We have even better photos of this now, such as is found in Figure 15, from the Hubble telescope. The image is worth pausing over: each galaxy shown here contains billions of stars. In reading the previous sentence and examining this photo, one can find oneself staring incomprehensibly at this picture: each of these circles contains billions of stars, you say?

Sagan often reflected on the relationship between large numbers and our understanding of ourselves and the cosmos. Of course he is famous for the phrase "billions and billions," which was actually coined in Johnny Carson's parody of Sagan. Sagan's book *Billions and Billions* embraced this phrase and reflected on the importance of using and contemplating such incomprehensible numbers. In discussing exponential notation (e.g., 10^{34}), Sagan notes that "this doesn't mean you can *picture* a billion or a quintillion objects in your head—nobody can. But, with the exponential notation, we can *think* about and calculate with such numbers."[66] Yet in an important sense the image below *is* a picture of billions and billions. But these scalar relations are deceptively concrete: together the accumulation of billions provides a reference point for just how the image ought to overwhelm. Sagan is thus following Humboldt when he suggests that these scalar images and numbers are ultimately an elaboration of the excess, the unending wealth of possible calculation, objects, and description possible even within those things that seem to be relatively apparent.

Sagan highlights how scientific descriptions contain a kind of compression factor that condenses an enormous accumulation of discoveries and ranges of size into something simple and deceptively concrete. In invoking any scientific description, it becomes possible to selectively switch scales to provide new modes of description, all in some way incomprehensible. Thus he gives us the beautiful scalar number: there are 10^{80} elementary particles in the Cosmos.[67] Such a

FIGURE 15: Islet on islet of scattered galaxies. Image by NASA; ESA; G. Illingworth, D. Magee, and P. Oesch, University of California, Santa Cruz; R. Bouwens, Leiden University; and the HUDF09 Team.

number is not meant to be comprehended; although it can be used for calculation, it is a notation of the excess beyond what any particular *Homo sapiens* can see and understand. Rather than expressing a frustration in the sight of this not-knowing, as we have undoubtedly felt when contemplating a number like the trillions of dollars in national debt, we might recognize the productive disorientation provided by such a notation. To do so is to direct ourselves to attend to the scalar reconfiguration.

As a mode of defining variables and examining their relations, science will always be able to proceed with these images and descriptions, since they will yield practical means of calculation and manipulation. But, as we saw in Humboldt, this elaboration retains the connection to excess, which manifests as bewilderment or wonder. To recognize this wonder is to tune in to how the seemingly complete depiction actually exceeds itself. To sit in this bewilderment is to unfold this diversity within the contours of experience and learn to rework our sense of reality within these multifarious, wondrous layers. May we all learn to sit in this diversity without fear, without excessive respect or disdain for these textbooks, fact checks, and journal articles, and with a patient, careful, yet exuberant wonder.

11

The Cosmos Seeing Itself

Representations of Scale, Scales of Representation

DEMANDING THE WHOLE EARTH

With his consciousness assisted by the bio-evolutionary adjunct named LSD, Stewart Brand gazes from the balcony of a high-rise. A vision emerges:

> The buildings were not parallel—because the earth curved under them, and me, and all of us; it closed on itself. I remembered that Buckminster Fuller had been harping on this at a recent lecture—that people perceived the earth as flat and infinite, and that that was the root of all their misbehavior. Now from my altitude of three stories and one hundred mikes, I could see that it was curved, think it, and finally feel it.[1]

This experience allows Brand to not only intellectually contemplate the roundness of the planet but also to extend his understanding beyond everyday perception so that he can finally feel what he had before only heard, inducing a monumental and counterintuitive reexamination of some basic assumptions that permit all kinds of "misbehavior." Immediately, the literally rhetorical question arises: how can we get others to transform their perspectives likewise? "A photograph would do it—a color photograph from space of the earth. There it would be for all to see, the earth complete, tiny, adrift, and no one would ever perceive things the same way." The hope here is that this photo would act as a nonchemical yet psychedelic—mind-manifesting—adjunct for consciousness, providing a view of a larger scale that would transform both our conception of this planet and our behavior on it.

This vision yielded the "Why haven't we seen a photograph of the whole Earth yet?" campaign, in which Brand sent badges with this question to a number of members of Congress, public figures, NASA officials, and even Buckminster Fuller and Marshall McLuhan.[2] When

Brand published the *Whole Earth Catalog* two years later (1968), he found just such a photo, taken from an unmanned satellite, to put on the cover. In 1972 the astronauts from Apollo 17 captured the Blue Marble photos, eleven pictures of the Earth in full phase. These photos have been called the most reproduced pictures in human history, and few will doubt their importance in launching the environmental movement and becoming a symbol for a globalized mentality.[3] Half a century later, however, such images are so commonplace that it is difficult to understand the excitement expressed by Brand.[4] What did he mean "no one would ever perceive things the same way"? How do we perceive things differently now that we have this view? Given that we all routinely use a similar image—if we'd like, rendered in beautiful interactive 3D—to find our way across town, it is difficult to see Brand's statements as anything more than a hyperbolic and self-indulgent (even drug-induced, we might say with that grin that those who grew up in the United States in the middle of the drug war find it hard to suppress) wish for a radical social change so prevalent in the counterculture of the 1960s.

Most responses to the photo written today are largely dismissive in this way, if not completely critical of the hope for transformation found in Brand's statement.[5] Such a critique is not unfamiliar, however, given that the magazine born from Brand's magnum opus, *The Whole Earth Review,* itself published such a critique, by Yaakov Jerome Garb, less than two decades following the release of these photos. In this reading we see a familiar turn to the idea of metaphor: "The image of the whole Earth—proudly displayed on the front cover banner of this magazine—is our culture's current metaphor for the Earth. This photographic image is not the reality of the whole Earth, but only one possible interpretation of it."[6] The article then examines "those qualities of the whole Earth image which most environmentalists *do not* seem to be aware of, and the ways in which this image is being used to cultivate attitudes that are *destroying* the Earth."[7] For Brand, however, this image is hardly as arbitrary as this argument implies. Rather, the vision is about a specific alteration of the way we understand ourselves and the cosmos. The image of the Earth, Brand hopes, will transform our view, since it will provide an experience of something that would otherwise be difficult to experience. The problem is that both these appropriations and the criticisms of them arise from the same maneuver: the whole Earth image is cut off from the organizing principle that grants it its power. That principle is, of course, scale.

This chapter is about the general confusion presented by scalar representations. Such representations are necessary for reorienting and rewriting our worldview according to new scales of existence. In examining these representations, cultural critics tend either to celebrate scaling (usually to the large) or to critique it; there has been less reflection on the nature of scale and how this clarifies the narratives, images, and other representations of scale.[8] The persistent problem for cultural critics is understanding the relationship between the origins and parameters of scale and the way we represent it. Since scalar perceptions are constructed out of a regulated and consistent accounting of the changes in observations, scale provides some standard for the function and form of scalar representations.

Much of the current scholarship on scalar representations remains critical of scale.[9] When scalar perspectives arise, many of the theories and ideas we have discussed thus far are frequently inserted as critiques of scale: Haraway's critique of the view from above, Cosgrove's colonial eye, and Arendt's concern about leaving the Earth are some of the most common.[10] We have already shown the limits and applicability of many of these critiques in previous chapters. Here I want to highlight why depictions of scale are so easily critiqued by explaining the means of their distortion. Doing so will make clear the interpretive challenge provided by scalar images and outline what is required for scalar images to have the kind of power Brand hopes for. The basic argument is simple: scalar depictions are only powerful if they are encountered as scalar. In turn, reinserting scale into such depictions often undoes or reworks many of the appropriations and interpretations that bring them the most criticism, while providing a standard for critiquing depictions according to the degree to which they work within a scalar framework.

SCALING BEYOND METAPHOR

Garb's critique of the Blue Marble relies on a strong conception of metaphor derived from the linguist George Lakoff. On this basis, he provides a series of critiques of the whole Earth image. Examining this metaphorical reading of the Blue Marble will make clear the structure of scalar representations.

If the image of the Earth from space is a metaphor—"a way of describing one entity in terms of another," as Lakoff and Garb both define it—then what are the two entities being described?[11] The image itself is on the same scale as our usual experience; it appears to be a clear entity.

But what is this other "entity" it is put in relation to? The difficulty is that this second entity is *incapable of being experienced* in a way that clearly defines it as an entity. Thus, the whole Earth image creates an experience for something that, barring our own personal space travel, we are otherwise incapable of experiencing: everything in our usual experience held together. The result makes concrete a scalar idea (the Earth) through a new scalar object (the Earth seen from space) re-presented in a way visible to these *Homo sapiens* at this scale. Thus, what scalar representations create is an experience about experience: that (the image) is how you experience all of this, when viewed together.

The presence of this scalar relation makes the metaphor non-arbitrary because of the consistent standard provided by scale (2.3) that we hold ourselves to (2.20). The most rhetorically powerful representations will then use a visual form that ties the view to a consistent empirical, human-bound sensory apparatus. In this form, the consistency of the apparatus noted in 1.11 and 2.6 is combined with the measure of scale in order to make both experiences present and adequately situated in relation to each other. The resulting image re-presents this scalar relation as a visual experience, but in order for it to be appropriately understood, the scalar relation must be apparent. To put it in more potent terms: the second entity in the metaphorical relation is *normalized human experience itself,* which is then transformed by the view captured in the image (2.16). This seems to be what Brand desired from the image: something to render concrete this scalar relation that pulls us outside of our normal experience and renders it on a new scale, while still retaining the persuasive power of the empirical. Literally, you see it.

But we are only affected by it if we include the scalar relation embedded within the metaphorical relation. We can see now why this image presents a great risk: in making something outside of usual experience look like something within usual experience, we risk treating it like just another nonscalar object. The scalar relation becomes buried and, with it, the reference back to and revision of your normal lifeworld. Just as science encodes this relationship into its descriptions, we can lose sight of the scalar relation in the apparent object of representation. For Garb, this elision manifests first as the statement of the priority of the metaphor through the following epigraph from Lakoff:

> Metaphors may create realities for us. . . . A metaphor may thus be a guide for future action. Such actions will, of course, fit the metaphor. This will,

> in turn, reinforce the power of the metaphor to make experience coherent. In this sense metaphors can be self-fulfilling prophecies. . . . The way we think, what we experience, and what we do every day is very much a matter of metaphor.[12]

This prioritization of metaphor permits Garb to treat the image as both the cause and the manifestation of certain ways of thinking. But the central organizing mode of thinking here is not derived from a metaphorical relation but a scalar one. The elision of the scalar relationship permits Garb to attach other interpretations to the image and argue that these readings are themselves *in the image* rather than in ways of approaching the image. Thus he warns that "this banner of perspective and insight is also—just below the surface—a banner of alienation and escape from the Earth."[13] Rather than "just below the surface," the readings presented by Garb are in some sense on the surface, in that they treat the image as just another object to be arbitrarily manipulated. Such readings are not really about the image at all but rather the way the image becomes reworked into different assumptions, interpretations, and uses that are more or less cut off from this scalar relation. Consider the result of Garb's reading. The bulk of his article consists of objections that are all critiques commonly repeated about the view from above: it is a privileged view, emphasizes the "out there" over the here, is a presumptuous "God's-eye view," gives up on the Earth as our home, produces a homogenizing understanding of the diversity on the planet, allows us to neglect the being in "nature" that we feel on the surface, contains the Earth in a single object, becomes an object of domination, and ultimately allows the Earth to be further trivialized and exploited. Undoubtedly, the image, treated just as a metaphor, could be used to these ends or read in these ways. However, such uses treat the image as just another object to be manipulated; doing so, they fail to attend to or distort the scalar relationship behind the metaphorical one.

Garb's concern over these alternative uses of the image arose from what he saw as a defilement of it. In a small account found in the sidebar at the beginning of the article, Kevin Kelly recounts a conversation in which Garb flips through a series of appropriations and remixes of the whole Earth image. Frustrated over the defilement of this "totem," Garb goes on to write this essay.[14] This feeling is not unusual but arises anywhere there are totems for a simple reason: they have power. And anywhere there is an image of power, some will try to take advantage of it by making use of such surface readings while avoiding the truly potent elements contained within it. If we reinsert the scalar relationship, we can see that such appropriations are actually scalar distortions. The critical task at hand, then, is to rediscover and reinsert scale as the underlying transformative perspective that gives this image its power.

SCALE CRITIQUE: RESPONDING TO SCALE TRICKS

Because scalar perspectives can only be encountered through some representation, and because they are so counter to our usual way of experiencing the world, we find ourselves with all kinds of distortions, tricks, and erroneous depictions of scale. I want to pause here to discuss four kinds of scale tricks that lay the groundwork for what Derek Woods has already named "scale critique."[15]

The first is the assumption that something might get larger or smaller indefinitely. It is partially to target this kind of distortion that we, at 1.31, separated Gulliver's scaling from the kind of scaling we are dealing with here. The essential difference is that Gulliver's scaling takes given objects and makes them larger or smaller, whether in actual scaling of operations (the growth of an elephant, tree, or business) or in representations (e.g., Godzilla or King Kong). These kinds of scale are usually conflated, because discussions of scale often start with an object in view; we then consider how such things get larger or smaller. The distinction highlights the limits to this scaling of objects, particularly for living systems (see 3.29). Distortions of these limits can be relatively harmless and even educational (e.g., the classic trope of shrinking small enough to go into the body). However, they do not provide good training in scalar thinking. More importantly, they indulge the tendency to think of things as capable of extending their reach indefinitely despite the inevitability of scalar thresholds.

This notion that growth might expand indefinitely, especially in an economic sense, has been extensively critiqued, most recently by the anthropologist Anna Tsing.[16] Tsing notes that the "term 'scalability' had its original home not in technology but in business. Scalability in business is the ability of a firm to expand without changing the nature of what it does."[17] Tsing rightly examines the tension between ecological constraints and economic fantasies of unlimited expansion, which she situates in the unintended effects of agricultural expansion.[18] At the same time, her arguments about nonscalability misrepresent the problem as an issue with the precision of scale itself. The problem is not the "precision-nested scales" but rather a fantasy about the indefinite scaling of objects that is itself undone by carefully attending to scale (3.29).[19]

The second scale trick is the overlaying of two very different perspectives without a scalar notation marking this difference. In 2.19 we noted that one way of making these distortions in film is to overlay two perspectives in the same field of vision. This creates a distortion, since the eye does not have a way to distinguish the two perspectives and therefore treats them as one. In this technique we have a clear image of how we perform this overlaying in many images, descriptions, and terms, particularly those that bridge two scale domains. For instance, in 3.32–33 we noted that "all humans" is a term that bridges two scales; humans are only visible at the meter scale, yet this grouping pulls our perspective to the ecological. As we noted at 3.24, these maneuvers are not always a problem. The problem arises when the connection between scale domains is forgotten. For example, take the term *Anthropocene*, which has recently become a rallying point for thinking about the effect of humans on the planet: like the conflated perspectives in film, this term wraps together a this-scale object (humans) with a planet-scale one (geological patterns) in both space and time. Many writers on the Anthropocene are attempting to productively separate these two notions of the human. Thus, Dipesh Chakrabarty's four theses on the Anthropocene clarify the scalar nature of this term:

> We can become geological agents only historically and collectively, that is, when we have reached numbers and invented technologies that are on a scale large enough to have an impact on the planet itself. To call ourselves geological agents is to attribute to us a force on the same scale as that released at other times when there has been a mass extinction of species.[20]

Chakrabarty acknowledges the differences in scale that make planetary effects possible. The tie between "Anthropocene" and the human is thus essential for highlighting the relation between the two different scales of interaction.[21] At the same time, the ambiguity of the "us" for the attribution of agency and the work performed "historically and collectively" creates a potential scalar conflation that manifests as the double binds of environmentalism that we noted in chapter 9: we both are and are not the agents. But this is also why Woods has highlighted scale variance—the way objects change across scales—as essential for scale critique: to get from humans to the Anthropocene, one must first cross a scale domain where this phenomenon "human" now becomes something different.[22]

The third scale trick is forgetting the scale you (or any given action) are on. As we saw in 1.14, the human body does not change scales even in scalar perceptions such as the view from space. In a more general sense, we always risk mistaking these scalar extensions and perceptions as a displacement of the capacities tied to these bodies. We generalized this in 3.34 in the phrase "one can only act on the scale on which one exists." We have seen a potent conceptual version of this in Thomas Nagel's distorted view from nowhere in chapter 8. Likewise, in chapter 9 we discussed Erik Drexler's fantasies of extension as an example of misplacing the scale of the engineer. These problems naturally arise as we learn to develop and work with scalar extensions and scalar relations. We can follow Timothy Clarke and Joanna Zylinska in calling these "scalar derangements," since they are distorted fantasies of extension to and intervention on other scales, while forgetting both where one is acting from and what scale requires for such action. Thus Clarke's examples of derangements of scale include the "you control climate change" trope.[23] Such rhetoric avoids the most difficult aspect of ecology: the need to rethink the here in relation to new data about collective endeavors without reducing one to the other.[24]

A fourth object of scale critique is scalar overwriting, in which concerns from one scale (usually the meter scale and filtered according to human needs) are mapped onto another scale. We are familiar with this overwriting in mapping software such as Google Earth, which makes it easy to draw human points of interest onto the surface of the Earth. This operation is standard practice in mapmaking; this is one reason why we distinguished cartographic scales from scale itself at 1.32. The major risk here is that such overwriting permits you to retain the priorities and points of interest from another scale. Doing so can be essential for scalar reorientation, since it assists us in tracing elements relevant for another scale, such as describing DNA in terms of what it might produce at the level of the body or when a particular bacteria or toxin produces organism-level effects (as was performed in 6.14–17). However, in such overwriting we need to attend to which terms apply to which scale and how; otherwise we end up with one of the distortions just discussed.

Combining this mapping with the above scalar distortions, critiques of satellite imagery and similar scalar cartographies bring our attention to how views of different scales can be accompanied by attempts to overwrite and claim such scales. For example, the architect Laura Kurgan's extensive critique of satellite imagery traces the way that such devices are put into service for various ends, from the domestic to the military.[25] Such critiques trace out the conditions for the use and production of these images so that we can test claims such as whether "high-resolution satellites . . . signify global transparency."[26] Importantly, the abuses of this ability to visualize are often a result of exaggerated claims reliant on scale tricks and distortions. Thus, Kurgan notes, in a way that mirrors discussions of visualizing nanotechnology or microbiology, that "no satellite image presents a simple, unambiguous picture of the Earth, and a visit to the site itself can often raise more questions than it answers."[27]

In all four instances, scale critique would be the process of comparing these representations of scale with how scale functions in practice. The notion of scale tricks or distortions requires (1) scale as a consistent standard, (2) tying together the two perceptions, and (3) locating the perspectives, including the one that is hidden (usually in the viewer). However, this operation conflicts with a common tenet of cultural criticism, since it requires a reference to a standard. Thus, Tsing and Woods both take issue with the notion of consistency and precision in scale. To be clear, scale does not require a return to a truth standard of representation for a world taken as given in itself, nor is it a tyranny of precision. The precision here is based on the standard implemented by scale, which produces a consistent means of accounting for our observational shifts. Because scale is about observation, the standard of *scale is a holding ourselves accountable for these shifts in observation* (2.20). This is why scale is so powerful for critique: scale's simple yet precise accounting for these changes in observation provides a standard for critiquing distortions in our perception of reality and the way we impose on scalar views assumptions derived from a nonscalar, human-based lifeworld.

PORTRAYING SCALE ITSELF: THE POWER OF SCALE IN *POWERS OF TEN*

While scale critique would use scale to critique scalar representations, the task of portraying scale itself is a separate endeavor. This task is inherently fraught, because there is no comparable phenomenon that appropriately isolates the way that objects transform into other objects in changing scales. *There is no metaphor for scale that can be drawn from our nonscalar lifeworld.* The common metaphors for scale are all derived from this-scale experience: concentric circles, the *matryoshka* doll, or the ladder.[28] Therefore, their modes of embeddedness and layering do not capture the essential reconfiguration presented by scalar experience.

The most adequate way of isolating and demonstrating the power of scale is to take an image of this-scale experience and show it transforming into other scales. To this end, the paradigmatic scalar depiction is Charles and Ray Eameses' 1977 short film, *Powers of Ten*.[29] The concept of *Powers of Ten* is relatively simple: it begins with a view of two picnickers sitting in a park and then alters the frame of reference, increasing the scale by a factor of ten every ten seconds until it reaches 10^{24} meters.[30] It then does the same back down in scale until 10^{-16} meters, the size of subatomic particles. In this movement—from a normal view to a planet to a solar system to galaxies to groups of galaxies and then back down to cells, molecules, atoms, and subatomic particles—we see not only the various objects at these scales but also the apparatus of scale itself.

This film is often cited in critiques of scale, with almost universal criticism.[31] Woods captures the general sentiment when he argues that "if any film qualifies as an example of what Donna Haraway calls the god's eye 'view from nowhere,' *Powers of Ten* is that film."[32] Our account of scale will assist us in responding to the three major critiques. First, almost all of these critiques focus on the seemingly "smooth zoom" produced by the film (see 6.6). The film produces the illusion of movement, and yet it is not, as Vittoria Di Palma argues, moving at all but is rather a change in resolution.[33] The result for Di Palma is that "the zoom erases divisions between scales, and its rapid progression from one extreme to another renders space and distance irrelevant."[34] Second, critics highlight the artificiality of the film's transitions between scales, as is best characterized by Latour's question, "In what laboratory would we have to visit to observe cells from the skin of our two amorous picnickers?"[35] Third, many critics argue that the film preserves an anthropocentric and mastering perspective for the viewer (the essence of Woods's invocation of it as a god trick).[36]

We can confirm that such a zoom is a change in resolution, which became a key concept for scale here at 1.15 and throughout Part I. But the idea that this smooth zoom erases the divisions between scales ignores an essential aspect of the depiction: the film shows the changes in scale domains through the transformation of the objects being seen. Space and time are not irrelevant so much as reworked in the logarithmic shift. In response to Latour's critique about the constructed nature of the images, we can say that this is precisely why we need a depiction like *Powers of Ten:* scientists do not usually patch together the diverse scales they are mapping. Rather, sciences like ecology, geology, and meteorology produce knowledge at the planetary scale by observing differences that are discernible at the level of the planet. It is because these knowledges are so disjointed that we need something to tie these disparate scales of observation together to portray scale itself. The smooth zoom shows how these levels relate together, using the continuity of the scalar shift to visualize the transformation of our lifeworld into these other scales.[37] This is what Woods describes as "an explicit effort to describe new objects of science and alter the phenomenological worlds of its viewer."[38]

Woods, however, denies *Powers of Ten* this status, since it "domesticates the uncanny or paradoxical activity of nonhuman scales using Protagorean/Vitruvian measurement norms of the human lifeworld."[39] Interestingly, Monica M. Brannon makes almost the opposite conclusion in her reading of the film, stating that "this new standardized norm conditions a perception that privileges a mechanical objectivity over the narrative, on-the-ground social experience; this way of seeing ultimately reduces spaces to renderings of mathematical measurements."[40] Woods and Brannon can make opposite points about this same aspect, because scale is a consistent extension of a measure that is derived from human experience (2.20). Brannon's critique is not accounting for how this measure is actually tied to "on-the-ground" experience through its reference to an object (e.g., a meter) derived from human experience (2.10, 2.14). Woods's critique is not accounting for how this extension of the measure pulls experience into these new phenomenal realms by ensuring that scalar experience is tied to and therefore transforms our understanding of our this-scale lifeworld (2.11).

In retaining a reference to our normal experience, in using the consistency of mathematical extension to move beyond this experience, and in projecting out a view of the world in the familiar terms and format (i.e., visual images), scale ties these transformations of the cosmos to ourselves. Scale is a holding-ourselves to the consistency of a measure derived from our own experience (2.20) and in the process, as we saw in chapter 8, transforms what we think we are. In this light we can be sympathetic to critiques of the narrator of *Powers of Ten,* since the presence of this voice—particularly the attempt to fill up the less interesting parts of the exponential climb—could work against the transformation being performed by the visuals. Perhaps, as Woods argues, "in *Powers of Ten,* the narrator maintains his identity on this vast trans-scalar journey."[41] Perhaps we could do so as well, leaving our own lifeworld intact and relatively untouched by the images we're seeing. To do so would be to externalize the whole representation as being simply a representation. If, however, we actually take the transformations being experienced as a transformation of our own lifeworld, then the radical reorientation of scale manifests.

LOSING SCALE, LOSING OURSELVES

In contrast, depictions that attempt to cut scale out of the picture lead toward rather different effects. For example, an online graphic by the artist Josh Worth, called *If the Moon Were Only 1 Pixel: A Tediously Accurate Scale Model of the Solar System*, is built on a refusal to switch scales. It sets its scale at the size of the moon as one pixel (hence the title) and then maintains a continuous scale. The result is a line that allows the user to scroll endlessly through what appears to be nothing. The graphic is valuable for showing how we elide the significant shifts in scale required for understanding the immensity of space. However, this graphic does not actually show scale itself, but rather a single other scale. Not content to merely leave this lesson stand on its own, Worth includes some musing on the emptiness that one can discover at this scale: "When it comes to things like the age of the earth, the number of snowflakes in Siberia, the national debt. Those things are too much for our brains to handle. [after some scrolling] We need to reduce things down to something we can see or experience directly in order to understand them. [more scrolling] We're always trying to come up with metaphors for big numbers. Even so, they never seem to work." While we can agree with Worth's statement about the incapacity to formulate metaphors for scale, the negativity of this sentiment arises from the fact that the graphic is not providing a picture of scale itself but a picture of one single scale. Worth seems to miss that the value of such orders of magnitude lies in how they make these larger things at least marginally comprehensible.

Furthermore, because he is staying on the same scale, Worth is not showing the things already present between and among these large-scale objects. In this elision he finds himself (as we continue to tediously scroll through, feeling ourselves a little bored and listless) able to posit:

> It's a good thing we have these tiny stars and planets, otherwise we'd have no point of reference at all. . . . All this emptiness really could drive you

> nuts. . . . The brain isn't built to handle "empty." "Sorry, Humanity," says Evolution. "What with all the jaguars trying to eat you, the parasites in your fur, and the never-ending need for a decent steak, I was a little busy. I didn't exactly have time to come up with a way to conceive of vast stretches of nothingness."

But evolution did equip us with the capacity to feel wonder at vastness and interconnection. Worth's graphic cuts out both of these elements by refusing scale and refusing to include those things that do actually fill this nothing, but which he can't show because he's not changing scales. The result is boredom, separation, and lack of meaning. This selective scaling also permits him to project this nothingness down to the scale of atoms as well:

> Emptiness is actually everywhere. It's something like 99.9999999999999 99999958% of the known universe. Even an atom is mostly empty space. . . . Some theories say all this emptiness is actually full of energy or dark matter and that nothing can truly be empty . . . but come on, only ordinary matter has any meaning for us.

This statement privileges the apparent location of the stars and ignores the fact that every star is sending an array of electromagnetic rays and gravitational fields that extend across the entire cosmos. But, as Joel Primack and Nancy Abrams note in their work on cosmology, all things are in some way insignificant on other scales.[42] From a fully scalar perspective, such insignificance is positioned in the larger schema of interconnection we elaborated on in chapter 7. The nihilism and glorification of this kind of emptiness is built on a persistent exclusion only possible if we ignore scale itself.[43]

BEHOLD THE EARTH

What Brand recognized, standing on his balcony and experiencing the curvature of the Earth, was that new modes of imagining and imaging were now possible for making present these seemingly inaccessible scales. Only in the twentieth century can we actually arrive at a physical position where the Earth might be viewed with the same visual apparatus that is used at this scale. While very few of us have made it to that physical location, visual technologies allow us to vicariously take our vision to this perspective in a way more akin to our possible experiences of smaller scales. This movement constitutes a fundamental shift in the rhetorical weight of the images and descriptions produced. However we would like to critique both vision and film technology, these photos and astronaut accounts are on equal grounds with the claims already made by all of science grounded in empiricism.[44] This is what the Earth looks like, we can say—not naively but in recognition that insofar as we want to invoke any authority of the empirical, this photo also claims these grounds. I thus want to end this chapter with an encounter with this image without falling into the clichés that characterized the rhetoric of space travel around the Apollo missions, and which works through the trivialization of the familiar.[45] We can thus turn now to the photo itself (Figure 16) to rediscover its capacity to induce scalar experience.

The disc is surprisingly bright—luminescent on my screen, a swirling of white and blue. I feel almost immediately the force of my usual orientation as I recognize Africa and the Arabian Peninsula. Glimmers of Asia appear on the top edges. I am looking directly into the gap between Africa and Madagascar—or slightly below it where the photo appears to be a bit brighter. The rest is ocean and cloud, although after a minute I realize that the big patch of white on the bottom of the disc is probably Antarctica. It appears as if the whole planet is reeling back away from me. But this is my normalization of the planet's image withdrawing as my brain attempts to orient itself toward the image in a familiar way. These names, orientations, and shapes arise from the thousands of times that I have seen these areas mapped, overwritten with names and borders and places. I am so saturated already with weather maps, geography lessons, and forays into Google Earth that my mind overlays it with those yellow lines and little names. I almost want to reach out and spin it, to search for the place where I am or would like to go.

But none of that is actually here in this photo before me. The longer I look, the less familiar it becomes. That strange textured brown and dark green slab is a strange exception to the white and blue hues that intermingle organically across the disc. The shape of it is no longer what I know as "Africa"—I can't even see the edge of the landmass on the top left; I was adding it in, knowing it would end there. The longer I gaze, however, the more I have the feeling that even if this globe was oriented the way it often is now—North America or Europe in the center with each continent clearly distinguished and with less emphasis on the ocean—I would still be able to let go of that mapping. These are, after all, only the shapes of landmasses.

FIGURE 16: The Blue Marble. Image by NASA.

When I stare for a while at the bottom right section, even the designation of cloud and ocean drops away as my mind switches duck-rabbit style between the two, unsure which one is the surface and which one is not. When I let this gaze fall into the division between land and water, the continuity of the clouds produces a surprising depth. Feeling this depth, I let my perspective widen to take in the whole picture. As an experiment, I back away so that I can take it in fully at once. Apparent orbness manifests, the reality of the curvature that Brand glimpsed from his balcony. The edge is suddenly magnificently apparent. My eyes follow it around for several rotations, delighting in the contained contrast with the stark black.

A consideration arises—no, a feeling invades my senses: what *am* I looking at? I look away for a moment, consider my desk, the scattered books, the light from the window, the trees outside blowing in the wind, my chair, my computer, my cup of tea—where is all of this? Even if this photo was centered on me here, in that splotch of brown and green shaped like the form I've come to know as "North America," the truth is *I perceive none of it.* Or rather I know I perceive it, but—I seek for another means to say it but can only go to a scalar comparison to describe the experience—it is like saying I see the atoms in my hand when I look at my hand. My mind reels at the suggestion, experiencing both scales at once for a moment. *All of this here, is in that orb.* And this orb is seamless, vibrant and clear in its spherical vastness. Now the duck-rabbit switch is my existence itself, felt whole, divided, whole, divided. In this disorientation a new thought emerges, as if the voice of the Earth itself: *In this multiplicity, the veritable detritus of matter scattered about your little room, do I contradict myself by showing myself as a single object? Very well then, I contradict myself: I am large. I contain multitudes.*

Whitman echoes in my head as I bring up these lines on my second screen and see: "I concentrate toward them that are nigh. I wait on the door slab." The picture is still on my main screen, waiting, demanding its vastness. I become aware of it replicated on the desk before me, on a book, and another, and another, and another—even a business card someone found from a religious organization and passed along to me. This image spins into an icon, each replication a kind of prayer to the vast capacity of this orb to contain such a myriad of beings even as it obliterates the distinctions and reveals itself whole. I feel suddenly as if I speak for the Earth, like my hands type for it, only thus as a kind of plea: "Listener up there! What have you to confide to me? Look in my face while I snuff the sidle of evening." I can no longer tell if I am seeing the image or speaking to it—I sidle around its surface—I feel its emanation and feel how I emanate from it. I feel lost in its surface but still embedded, strikingly aware of my feet on the ground, of the pull of gravity—that testament of the immensity of this presence. My mind wraps around this sphere, is contained in it, by it, for it. I speak, the Earth speaks, Whitman speaks (across time):

> I depart as air, I shake my white locks at the runaway sun,
> I effuse my flesh in eddies, and drift it in lacy jags.
> I bequeath myself to the dirt to grow from the grass I love,
> If you want me again look for me under your boot-soles.
> You will hardly know who I am or what I mean,
> But I shall be good health to you nevertheless,
> And filter and fibre your blood.

Failing to fetch me at first keep encouraged,
Missing me one place search another,
I stop somewhere waiting for you.[46]

Like the first chirp of a bird, heard through a waking breath, a touch of *agapē*—the effluent *ānanda* (bliss)—penetrates my being and drops into the sight of the Earth as a Whole.

12

Transformations by Involution

The Contemplative Practices of Scale, Scaling Contemplation

STEP TO THE RIGHT!

While undergoing a major stroke, the neuroscientist Jill Bolte Taylor experienced her sense of separation and egoic identification shut down. In the TED talk that popularized her experience, she describes the feeling in scalar terms: "I can't define where I begin and where I end, because the atoms and the molecules of my arm blended with the atoms and molecules of the wall. And all I could detect was this energy."[1] In recounting this transformation in perspective, the neuroscientist makes a strange statement: "How many brain scientists have the opportunity to study their own brain from the inside out?" Suddenly, the brain scientist realizes that it is possible to study brain configurations through experience. In this account, this possibility remains only an exceptional moment, as if studying the brain in this way is only possible in a stroke or similar event. Yet this maneuver brings to our attention the fact that experience itself is already a kind of study of the brain's neuronal configuration, only on a different scale.

When Taylor physically brings a human brain onto the stage, she creates a confusing contrast to the poetic description of the experience. It seems that the authority of the account lies in the material invocation of the brain and her training as a neuroscientist, yet the science here is general, drawing on widely misunderstood tropes about the left and right brain and reliant on somewhat blunt computer metaphors. In contrast, her description of the experience itself draws from a recognizably spiritual or even psychedelic rhetorical repertoire. Even if Taylor's scientific understanding is more elaborate and nuanced than she presents there, one gets the sense that these nuances would be less important; after all, this studying from the "inside out" is actually about the *experience* of these differences in brain configuration. The brain structure itself—especially the brain as a nonscalar object displayed on the stage—is less precise here than the description of the experience.

Taylor's account highlights the much-obscured scalar shift between experience (the scale of here) and the scalar views we often permit to stand in for this here. It also highlights an essential difficulty that is exacerbated by Taylor's inability to bridge the scientific with the part-contemplative, part-self-help rhetoric she deploys. This slippage from the contemplative to the self-help emerges precisely where Taylor reasserts familiar humanist tropes of freedom and choice: "And I pictured a world filled with beautiful, peaceful, compassionate, loving people who knew that they could come to this space at any time. And that they could purposely choose to step to the right of their left hemispheres—and find this peace." The basic sentiment here is fine enough. Yet in asserting that this operation is something "they could purposely choose," Taylor throws us into an inherently difficult contradiction, which we can use to divide self-help rhetoric from the more difficult contemplative discourses: if the transformation requires the eradication of the egoic self, then how can this be self-help, as a choice by a self kept intact? After all, Taylor's experience was not deliberately chosen in this manner but was induced by a stroke. This puts Taylor's experience in the same rhetorical dilemma as the psychedelic experience of ego death.[2] Both the stroke and psychedelic experiences may, in certain configurations, induce a productive and healthy experience of unity and ego death. But as Taylor's account reiterates, this experience does not convey what the nature of the transformation entailed, what the experience is, or, most important, the conditions for finding, navigating, and working with this shift *not as a function of a brain obstruction but as a transformation expressed at the scale of experience.*

The difficulty can be described as a scalar shift from noticing a physical change in the brain—or inducing it forcibly via chemical alterations—to observing and reacting to a change in your configuration of experience. The idea that one could study the brain from the inside out identifies an information disparity between the scale of the neuron and the scale of experience. An image of the 100 trillion neuronal connections would provide a picture of a physical object that is already information-poor in that it does not tell us what that configuration is an experience *of.* Likewise, inducing the experience via a stroke or drug might create a momentary neuronal configuration (such as an experience of wholeness or ego death or cosmic consciousness), but this momentary experience does not necessarily reorient the brain—these 100 trillion neuronal connections—to this experience that is about its configuration. The chemicals clear away or the brain heals from the stroke, the moment of insight ends, and the previous brain configuration may resume. Meanwhile, those who would prefer not to have a stroke or enter the mired landscape of post-drug-war psychedelic culture are left wondering about the real value and nature of this experience.

There is thus something of a cross-scalar dilemma that is poorly navigated by this "choose to step to the right of their left hemispheres," as if the operation is simply a matter of switching brain halves like a scripted move in a line dance. Although we can experiment here with what neuroscientific data might tell us—how it might help us understand this reconfiguration—something more appropriately scalar is needed to comprehend such transformations of experience. This scalar situation brings into view a fundamental difficulty with our conclusion in chapter 8: ego death, the death of the humanist subject, and the scalar reconfiguration of the *Homo sapiens* as a function of the Cosmos requires an incredible transformation of thought and self-conception. But the very conditions of this transformation cannot be placed back within this framework of humanistic self-choice, but rather must equally take into account the scalar dynamics of the transformation: "you" are not reorienting just "yourself" but also reorganizing a complex and not-divided system to attune itself to its indivisibility.

This is the task presented by the contemplative who, upon experiencing a glimpse of this undivided, non-egoic view, must nonetheless struggle to reorient to this new view. As Indian philosopher Sri Aurobindo notes, "Even when we have arrived at some glimpse of this idea" of the scalar relationship between the individual and the Whole "or succeeded in fixing it in our consciousness . . . it is difficult for us in our outward parts and active nature to square accounts between this universal standpoint and the claims of our personal opinion, our personal will, our personal emotion and desire."[3] The revision of reality implied by even the simplest of scalar shifts demands such a radical transformation that additional work must be undertaken to reorient "ourselves"—this complex conceptual-perceptual unit consisting of some 100 trillion neuronal connections, some 40 trillion cells, a host of ecological delimitations and patterns formed from billions of years of evolutionary alterations, and an engrained social and linguistic orientation—somehow this experiential coming-together must be reoriented into this new interpretive and responsive configuration presented by scale. But the momentum of this normalized *Homo sapiens* ego-centered view is naturally immense; "The hold of this ego-consciousness upon our whole habit of existence is difficult to shake off."[4] It is this task that leads the mystic to the contemplative, to practices that are designed to redesign and reorient one's whole being to a new vantage point beyond and exceeding egoic forms of consciousness.

The articulation here does not fit neatly into the discursive domains of the religious, ethical, or scientific. Rather, continuing our discussion of rhetoric and mysticism in chapter 10, this scalar dilemma provides one site for thinking through a kind of involutionary rhetoric: practices geared toward transforming one's mental structure from the inside out.[5] While the rhetorical tradition has primarily focused on convincing other people, this rhetorical task is a matter of persuading oneself—or what one thinks one is—of a radically different configuration toward existence. Taylor's account is problematic not because her experience was not empirically valid or scientifically rigorous but because it lacks rhetorical sensibility; it does not attend to the conditions under which the experience was found, the structure of the reflection she is performing, or how this communication might induce the same experience in others. Thus, we can look here at more coherent articulations that synthesize such empirical and scientific language with a more careful attending to the structure of the experience and the conditions for the transformation.

The task of this chapter is to consider the rhetorical difficulty inherent in this scalar reconfiguration as it manifests at the intersection of contemplative writing and this contemporary scalar encounter with the brain. Doing so will also permit us to describe a theory of reflection on scalar grounds so that we can articulate why scale requires this kind of practice, what scalar reflection is, how contemplative practices provide directions for this scalar reflection, and how the scalar discourses of science develop a more nuanced account of this involutionary maneuver. At the same time, we will have to handle certain objections to this whole operation, keeping in mind that these objections usually emerge from forms of *humanist reflection* of the sort targeted by Cary Wolfe's posthumanism, which attempts to distinguish itself "from the reflection and introspection associated with the critical subject of humanism."[6] As discussed in chapter 8, the critical subject of humanism cannot exist in the usual way in a scalar account, but this does not mean that there are no forms of reflection not based on humanistic grounds. Likewise, the whole possibility of reflection hinges on a critique of a persistent difficulty in philosophy and critical theory: whether or not one can overcome, remove, or even escape prejudice, language, or culture. Our scalar account will show how a form of scalar reflection can resituate our relationship to both physical and socio-cultural-linguistic scalar correlates to experience. Doing so will also provide new avenues for a theory of rhetorical transformation based not on humanistic grounds of choice but rather definitively posthumanist or even transhumanist (in Franklin Merrell-Wolff's sense discussed in chapter 8) practices of self-transformation according to self-transcendence.

THIS IS THE BRAIN LOOKING AT ITSELF

While Taylor brings a human brain onto the stage, the more likely evidence from neuroscience today would be the image of the PET scan or the fMRI. The rhetorical weight of such images has led to a noticeable increase in both popular writing and transdisciplinary academic collaboration using this neuroscientific data and vivid images to make extensive claims about human behavior and experience. This has led to what Mellissa Littlefield and Jenell Johnson characterize as the "neuroscientific turn" in academic research, as the evidence from neuroscience becomes invoked in more and more contexts. Littlefield and Johnson describe three such uses of neuroscience: using neuroscience to comment on subjects "traditionally under the purview of the social science and humanities" (e.g., religious experience), using laboratory evidence to bolster humanistic studies (e.g., taking brain scans of people reading a novel), and using neuroscience as theory.[7] A scalar account of the brain encountering itself could do all three, noting how this scalar experience has neurological effects, studying scalar experiences (such as cosmic consciousness) with an fMRI, or invoking these neurological changes as evidence for theoretical shifts. However, all three tasks first run into a prior and primary question: how do we understand this scalar bridge being performed by the *Homo sapiens* looking at the brain?

When the neuroethicist Judy Illes notes that, with neuroscience, "the flow of information has become multidimensional, multidirectional, and . . . multimeaningful," we can offer one reason it has done so: it has become distinctively scalar.[8] In doing neuroscience, the brain—insofar as we want to characterize or localize experience in the brain—is looking back on itself as an object, aiming to understand a new scale (the neuronal) as it relates to experience and cognition. In doing so, the brain finds itself in a strange situation about how to interpret the encounter. In a brain scan, the brain is producing a new reference point for experiential formations. By now we should recognize this scalar formation as the dislocation into (at least) two places (1.14, 2.8): a cognitive activity is encountered as brain scan and as experiential cognition. As became apparent in Taylor's account, switching scales is not a replacement of one scale for the other. For this reason, neurological studies usually rely on some correlation between a particular experiential element *with* the neurological configuration found in the scan. However, a few elements of this scalar loop make the situation more nuanced than some kind of direct peering into the brain structure.

First, much of what we might consider part of our experience is not within our direct awareness. Neuroscience may permit us to dissect forms of cognition not easily discernible in the field of self-referential awareness, discerning cognitive operations going on within our seemingly coherent and whole cognitive operations. Rather than simply using the blunt label "unconscious," such cognitive operations may be delineated through experimental formations designed to tease out otherwise implicit operations. The results are often hard to understand, partially because these nuances already encounter the narratives or interpretive habits built *about* our experience. For example, the infamous 1985 Libet experiment, which has recently been bolstered by multiple more-nuanced experiments, showed that decisions are "made" in some way by the brain a half second prior to the recognition that the decision is made.[9] These experiments imply that what we call "decisions" do not coincide with a particular kind of intentional action as we usually conceive it. We should be more precise in understanding what this shows than, for example, Brian Massumi's declaration that Libet shows how "thought lags behind itself" or that "the half-second of thought-forming is forever lost in darkness."[10] More precisely, it demonstrates a scalar incongruity in the experiential *interpretation* of something like decision making. That is, just as there are trillions of complex cellular interactions going on *within* the experience and function of the body, there are already complex cognitive formations going on *within* something like thought or decision making. The implication is that our narrative about (or introspective awareness of) "our" decisions is the thing that lags behind.

This nuance points to how there remains, even in this dive into the nonnarrative cognitive processes, a correlate or relationship that can be observed within our experience. Such observing requires a more careful attending to the aspects of cognition and experience. With careful methods of observation, contemplatives often make similar claims, noting the nonlocality of thought and decisions by observing that these thoughts seem to emerge on their own. One can observe within experience how thoughts, memories, and decisions seem to just show up—where did they come from? Whose are they? Did "I" produce them? These questions require looking carefully, slowly, and honestly at how thoughts and decisions emerge within experience. I would argue the Libet experiments can only be understood with these kinds of observations; if we fail to observe how such an experiment is evident within normal experience, then we will go on operating under the presumption that decisions are wholly and uniquely made by something called "me." Yet on what grounds is this claim made?

In observing these differences, the brain of the neuroscientist—or your brain—is not usually the brain put in the fMRI. As Taylor's account makes clear, brain scientists are not usually studying both scales at once, since they are studying an experience as localized in another brain. Furthermore, the need to aggregate such data means that any given brain, including the neuroscientist's, will always be to some degree disassociated from the specificity of what is discovered. The results will thus not fully capture the nuances of the experiential field but rather patterns or tendencies of cognitive formations. Given this limit, one must interpret such neuroscience in terms more general than your particular experience. Even more broadly, the brain itself cannot be said to be the sole location of cognition or many of the other attributes we ascribe to the brain.[11] While undoubtedly enormous amounts of processing occur in these estimated 100 trillion neuronal connections, these already bear a relation to the 40 trillion cells distributed through the body and the orders of magnitude more complex number of molecules and chemical interactions churning through the body. Likewise, the distributed cognition that occurs through linguistic and cultural forms, writing and technological reference, and projected media (light, sound, molecules for scent, tactile gradients) that permit senses to function all contribute to the cognitive formation that then expresses not only as experience but as the culmination of a cognitive, affective, or physical response.[12] Within these limits, one can resituate somewhat the significance of the Libet studies. As one more recent study examining Libet's results notes, such studies show "that the outcome of a decision can be encoded in brain activity of prefrontal and parietal cortex."[13] These studies do not find the source of decisions or any particular explanation of decision; they simply identify that a decision had been "encoded" in the brain and that this decision did not correspond to the self-referential recognition of a decision being made.

Understanding this complex situation of the brain looking at itself permits us to address the rather glamorous claims that a neuronal change *causes* something like Taylor's experience or the psychedelic experience of ego-death. Of course, the aneurism and the chemical clearly relate to an experiential shift in both of these cases. But the idea that the neuronal and chemical reconfiguration is *the* basis for conscious states is a clear case of mistaking a scalar shift for a causal relation (see 6.27). Taylor's experience does not prove that, for example, *nirvana* is merely a physical phenomenon; it only shows that a change in the configuration of neurons likewise creates a shift in experience. At the same time, the fact that such states can be forcibly induced points to their possible validity, but only if we more precisely describe the nature of the brain-experience transformation and then carefully map what it means to navigate the transformation involved.

YOU PROBABLY THINK THIS BRAIN STRUCTURE IS ABOUT YOU

With these nuances in view, a full sense of the contingency involved, and an acknowledgment that future studies may change this picture, let us look at some of the recent neuroscience to see how one can describe this ego-death transformation. There has been a surprising confluence of findings in neuroscience that bear on this question of self-identification. The approach here thus does not have to be concerned with specifically religious experiences, nor do we need to start with the growing literature on contemplative practices and the brain. Instead, we can start with what happens when you have people do nothing at all.

Through extensive brain imaging, neuroscientists have isolated what they call the default mode network (DMN, sometimes default network [see Figure 17]).[14] It is "default" because it tends to activate when people stop doing tasks. This is how the DMN was first identified: it activated when brain-scan subjects were not doing anything.[15] Once it was identified, researchers began to study what caused the DMN to activate, finding that it activated under activities that contain a self-referential content: "tasks that encourage subjects toward internal mentation, including autobiographical memory, thinking about one's future, theory

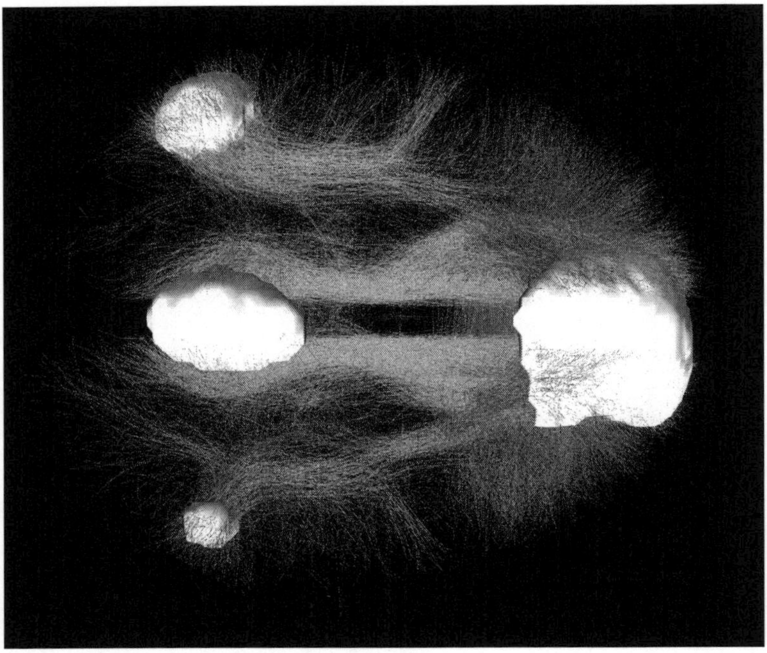

FIGURE 17: An image of the default mode currently in vogue among *Homo sapiens.* The irregular spheres here make up what neuroscientists have called the default mode network. The lines indicate connectivity between these nodes of activity. Default Mode Network Connectivity by Andreashorm is licensed under CC BY-SA 4.0.

of mind, self-referential and affective decision making."[16] These kinds of thought are said to be self-referential because they are about what you think you are, including a narrative about what you need to do or have done and a rehearsal of plans and memories. One study, which reviews thirty years of imaging, notes that "the default network is a specific, anatomically defined brain system preferentially active when individuals are not focused on the external environment."[17]

Many of these studies treat this DMN as an important part of an individual's cognitive well-being.[18] But, more recently, studies on meditation have independently demonstrated a correlation between meditative states and a decrease in DMN activity. One study scanned participants before and after several months of mindfulness meditation and found reduction in activity in self-referential parts of the brain.[19] Another scanned experienced meditators and found structural changes in the DMN that "supports the view that mindfulness training induces brain function changes that are accompanied by a reduction of self-referential thoughts during rest."[20] Similarly, Judson Brewer and colleagues scanned highly experienced meditators (over five thousand hours of meditation) and found decreases in DMN activity while meditating.[21] These studies suggest that meditative practices can work directly on this self-referential, self-narrative part of the brain, in line with experiential accounts of a withering of the self found in contemplative traditions.[22]

To add to this picture, studies unrelated to meditation found an anticorrelated relationship between the DMN and the parts of the brain primarily involved in high-level tasks, known as the task-positive network (TPN): when the TPN is activated, activity in the DMN tends to decrease.[23] These seem to suggest that the less the brain is taken up with self-referential thoughts, the better you are at doing high-level tasks. Brewer has recently applied this finding to his research on meditation, arguing that the relationship between the DMN and the TPN produces evidence for the need to release the hold one has on one's thoughts.[24]

Brewer and others have pointed out that a simple training can be done outside of any kind of meditative or spiritual practice in a cross-scalar encounter: if hooked up to an MRI, you could receive feedback about when the DMN is activated and attempt to adjust yourself accordingly.[25] Doing so could allow you to attend to this DMN activity, increasing focused attention over mind-wandering. But of course this loop only has more authority because we already struggle to attend to our own capacity to observe how we divide up the world. Indeed, meditation and other contemplative practices are already performing this loop as one observes the very processes of the mind as they emerge in experience.

This account may be tentative, and this interpretation has and will be contested. However, the wealth of experiential evidence, both in abnormal experiences such as Taylor's and in the accounts of contemplatives, already suggests that the basic conclusions are likely to be confirmed upon further research. This prediction might seem spurious or a means of generating self-validating conclusions (a significant problem for the science of meditation), but, much like the recent confirmation of the Libet study, the fact that so many people already attest to these experiential effects suggests that some neuronal reconfiguration is likely to be discerned within brain functions as the scalar correlate of these experiences.[26]

The suggestion here is *not* that the DMN is the I itself; that is, neuroscientists have not found your "self" in a part of the brain; they have merely identified a part of the brain that seems to pretend, that is, narrate to itself, that it is a self. The ego structure (see 2.24) is not an actual thing but rather a narrative told about what we are that gathers together a diversity of experiences and, in an act of scalar synecdoche (see chapter 9), calls them mine. In the end, the meaningful conclusion that can be derived from this increasingly specific scalar loop is this: the neurological shutting down of the sense of self implies that this sense of ego is a particular brain function, not some kind of deep-rooted identity. If "being separate" and "being an individual" manifests as a particular brain structure, then by implication that separateness and individuality are secondary distinctions created by a brain trying to understand its own experience. Suddenly, the materialist position through which one suggests that such experiences are simply bodily responses has been undone by this very locating of the self-referential, divisive narrative.

But proof is not what we're after; even if this declaration of the inessentiality of the self comes from a scientist rather than a Zen monk, it does not necessarily produce the transformation here in this brain. The brain looking at the neuroscience easily forgets that this is data about its own brain structure. The image it is looking at appears only to be an objective configuration; it can thus avoid the subjective, experiential configuration implied. In fact, we have to keep the subject-object distinction here precisely because of the way the results can be externalized. The problem is: in looking outward at these objects, the difference *could only be induced by an invasive reconfiguration* of the neuronal pattern—by something like a stroke or a drug. But then the experience appears to be caused by or require such an intervention. In recognizing the scalar loop, however, this brain examining itself can now turn away from the image and consider the structure of its own experience. Am I a coherent self? Who is this "I"?

WHOSE PREJUDICE IS THIS ANYWAY?

Before we describe the extracting of the habits of a divisive ego structure, we can revisit the opposite scalar correlate common within critical theory since the linguistic turn—the move to prejudice, language, or culture—as an additional means of externalizing the structure of cognition. One of the easiest means of dismissing contemplative experiences is to think of these effects in terms of prejudices, culturally bound interpretation, or psychological predispositions. Perhaps Edgar Mitchell was already predisposed toward the esoteric. Perhaps Philip K. Dick finally lost it. Perhaps, if Richard Maurice Bucke has been a bit more skeptical . . .

The academic argument around mysticism provides a site where this tension has played out, beginning in a debate between the philosophers W. T. Stace and Steven T. Katz.[27] Stace sought a basis for mysticism by defining features of what Robert K. C. Forman later called the "pure consciousness event." Stace claims that this experience is shared by all mystics but is articulated differently depending on the cultural and linguistic tools available.[28] In response, Katz argued, in a move recognizable after the linguistic turn, that all mystical experience must be inseparable from language and therefore reliant on the historical and cultural milieu.[29] Grace Jantzen later combined Katz's position with the articulation we saw from Haraway in chapter 8, arguing that "there are no views from nowhere, and there are no views from everywhere. There are only views from somewhere, and the particular place will have an inescapable effect on what can be seen."[30] Similarly, Richard King, invoking Hans-Georg Gadamer's argument that interpretation is foundational to being and that prejudice in interpretation is inevitable, claims that "there is no possibility of a universally applicable metanarrative or bird's-eye view of reality."[31] Further, King states that "the process of interpreting a text, then . . . inevitably involves the projections of one's own values, interest, and agenda onto the text."[32]

Our discussion of scalar experience presents three problems for this critique. First, scalar experience suggests a potential explanation for a pure consciousness event: it is a moment of moving past what one thinks one is and one's exclusive reliance on the reality of human-bound concepts. It is the brain realizing that it solidifies its own divisions—especially this division called the "self." As we saw throughout Part II, this reorientation is the arrival at the open position beyond a given set of divisions and values, the position from which new positions, judgments, and scales might emerge. To deny this possible state and to insist on the inevitable hold of language is to bind ourselves in the shadowy, ill-defined domain of the "human." But even scientific descriptions of scale indicate that this is not our situation. Second, this attempt to indelibly tie language or culture to experience is another example of a social counterpart to the physical claim that neuronal configurations cause experience. Third, and most importantly, if contemplative practice is already about examining structures of judgment—and pure consciousness events only manifest in or as this movement—then how can we ascribe the experience to these very things being looked at, given up, and transformed? We are thus led not to prejudice (in the sense of pre-given judgments) but to the *resistance to examining prejudice* that prevents us from rescaling our linguistic and social reference points. When we insist that linguistic and cultural aspects are inevitable and inescapable, we impose a content of the brain, a human conceptual framework, and a scalar delimitation of objects and identity back onto the experience that is about the undermining of these very contents. Content and prejudices are not left out of the account; they themselves are transformed in the process. Is this not the transformation needed as we recognize engrained structures of prejudice?

This maneuver of looking back at one's judgments is a persistent trope in contemplative rhetoric. For example, Augustine's transformation is described in terms of a scalar reflection with the major obstacle being his preexisting cognitive preferences. The whole operation required that he take into account his own mind: "Because I could not perceive them in the mind, I thought that I could not perceive my mind itself."[33] The problem is that Augustine does not know what his own psychical prejudices are: "I had no clear idea," he states again and again, "even of my own self" (119). The shape of this breaking down required a turning back upon his desires and thoughts. This turn back upon oneself is an essential trope—the Greek *tropos* itself meaning "turn"—by which the experience or realization is made possible. Thus, Augustine, upon hearing a story of another's conversion, states:

> As he spoke, you, O Lord, turned me back upon myself. You took me from behind my own back, where I had placed myself because I did not wish to look upon myself. . . . I looked, and I was filled with horror, but there was no place for me to flee to away from myself. If I tried to turn my gaze from myself . . . once again you placed me in front of myself, and thrust me before my own eyes. (156)

Augustine actively resists looking back at the aspects of himself that would become implicated in the conversion being demanded. Rather than a result of his prejudice or predisposition, this prejudice has to come into view as the very thing to be transformed.

The same applies with the scalar visions we have seen throughout this work. Because the vision is about what I think I am, I must allow this prejudice to come into view and have it reworked by this new perspective. Undoubtedly, it is difficult to implicate oneself into the scalar loop, to place one's prized conceptions, preferences, and prejudices within this extended measure. Given this difficulty, we can say that excessive skepticism tends to prevent individuals from allowing their attention to dwell on the scalar relationship long enough to process their own place in it. Augustine himself was, prior to conversion, one such skeptic. The difference hinges not on how critical or objective a person is but rather what one is critical and objective about: the structures of thought and habit within oneself. Augustine resists this whole operation: "I knew what it was, but I pretended not to; I refused to look at it and put it out of my memory" (156). This is a failure of being critical that masquerades as a criticism: the unwillingness to face his own structure of mind leads him to push it away, to suspect the apparent shift before him.

From this perspective, contemplative rhetoric often describes the operation of looking back at calcified cognitive structures as a surrender or realignment of will. Augustine, for instance, uses the language of linking formed by the investments of this body: "For in truth lust is made out of a perverse will, and when lust is served, it becomes habit, and when habit is not resisted, it becomes necessity. But such links, joined one to another, as it were—for this reason I have called it a chain—a harsh bondage held me fast" (151). We can read "lusts" here as the psychical investment in the things around us, the normal objects of value and importance, and the bodily needs that demand attention and addressing. Their importance becomes engrained in the scalar configuration of neurons, social mores, and conceptions of need that leads us to persistently reaffirm their structure and power in our everyday engagement with them. A different scale view contradicts, reworks, or reorients these linkages. Thus, Augustine is able to discern these two kinds of belief and emotional investment: those derived from the normalized human-confined structure and the larger, transformed values. Yet in this inward looking the conflict is only perpetuated by associating the difference with an ego structure: "It was *by me* that this habit had been made so warlike against me, since I had come willingly to this point where I now willed not" (151). When this war between the wills is released, the links themselves can be dissolved of their power. It was, however, not the belief that did so, but the surrender of it: "This was the sum of it: not to will what I willed and to will what you willed" (167).[34] The persistent surrender and transformation of one's beliefs, will, and prejudice are thus necessary to create a sustained transformation. An event of the sort exemplified by Taylor may start this process but be insufficient for a complete and ongoing reconfiguration of such engrained linkages.

Apart from this question about mystical experience, this account of prejudice is a way of rethinking the function of cultural theory more generally. Outside of the critical attitude diagnosed by Bruno Latour (see chapter 10), cultural critique can be seen as one means of tracing out the possibilities of humans configuring themselves in relation to reality. The purpose of such critique cannot be simply diagnostic but must become transformative: we must grapple with these remnants of erroneous and harmful ideas and confusions such as those called prejudice. To simply find these linkages in others, to identify them in our cultural narratives, or to say that we are bound by them risks voiding the whole operation from the start. Crucially, this is the whole terrain already mapped by contemplative rhetoric: contemplative rhetoric begins with the assumption that examining and transforming one's structure of mind is not some trivial endeavor but rather a difficult yet worthwhile process.

LOOPING THROUGH THE MÖBIUS STRIP

Our encounter with the brain above implies that the mechanism for this self-examination is already built into the structure of the brain itself. What is the capacity of the brain to consider itself? Because some of the confusion arises from mistaking the scale on which this view of prejudice occurs, I want to provide one diagram for this reflective process.

We can examine how this reflection works in a more rigorous way by once again using the language of cybernetics, specifically Gregory Bateson's short and provocative essay "The Logical Categories of Learning and Communication."[35] This essay approaches reflection from the perspective of learning more generally. Using the notion of logical types (see 3.6), Bateson is able to distinguish forms of learning by considering how a system refers back to itself (Figure 18). He begins by defining a base state where no learning is occurring. This Learning 0 (L0) might be considered the base activity of matter itself, which is what physics characterizes as action and reaction. Bateson suggests that the basic signals in information processing, such as sending a signal over a telephone wire, operate at this level because this communication does not

Logical Types of Learning

Learning 0 (L0)	Action/reaction	The base activity of matter itself (e.g., basic signals in information processing)
Learning 1 (L1) (learning about L0)	Stimulus/response	The base activity of living structures (e.g., Pavlovian conditioning)—provides basic "context markers" that signal an action to occur
Learning 2 (L2) (learning about L1)	Habit according to pattern of contexts	The organizational activity of complex living structures using models that organize contexts for responses (e.g., responses to people according to "character"). Reinforces itself according to self-validating demands of identity.
Learning 3 (L3) (learning about L2)	Control over or release from habitual necessity	Taking into account one's own context-making process without merely making it another context. To move beyond organizational contexts of habit (e.g., to reflect on one's prejudices in such a way that they are altered in the act of reflection).

FIGURE 18: Bateson's logical types of learning. Derived from Bateson, "The Logical Categories of Learning and Communication."

loop back on itself and therefore no "learning" occurs. Learning 1 (L1) is learning about L0. In L1 the organism provides basic "context markers" that signal an action to occur. Thus, in the iconic example of Pavlovian conditioning a dog learns to respond to a bell when it becomes a marker for food. We might say that L1 is the base activity of living structures since, in order to respond to its environment and maintain itself, any organism must organize its systemic responses in regular patterns.

Learning 2 (L2) is then the learning about L1. It occurs when a system uses models to organize markers for responses. For example, a bell might trigger salivation in the context of a restaurant but not in the context of a school. In this example, "restaurant" and "school" describe patterns of contexts markers. L2 includes most organizational activity of complex living structures. For instance, "airport" designates certain L1 responses as appropriate and others as not; if cued to stand on my head at the airport, I might not respond as I would at a yoga practice. Most human responses exist within linkages formed by L2, since we work according to organizations of contexts—complex forms of meaning, expectation, social situations, and abstract transactions. Something like prejudice begins on the level of L1, as simple stimulus-response, but becomes more manifest and embedded in L2, in which basic characteristics (he's black) become associated within an organization of contexts, such as capacities (he's not appropriate for this job), leading to particular responses according to that organization (he doesn't get hired). Most importantly, L2 is self-validating, since it tends to produce and reinforce the contexts under which it functions. Thus we replicate in our social, technological, or economic apparatuses the organization of contexts from which we function. As a result, "the self-validating characteristic of the content of L2 has the effect that such learning is almost ineradicable" (301).

Almost. Learning 3 (L3) is the learning by which we reflect on our means of making contexts and thereby learn to alter that structure. Such a structure, notes Bateson, is "likely to be difficult and rare even in human beings," and "it will also be difficult for scientists, who are only human, to imagine or describe this process" (301). The difficulty is that L3 requires one to take into account one's own context-making process without simply making this another kind of context. L3 cannot just be a change in contexts; this is the purview of L2. For example, to motivate yourself to exercise by setting a schedule, joining a running team, or getting a better pair of shoes is not a learning about your context-setting capacity but is merely an externalized changing of contexts in order to get your body to respond differently. Similarly, L2 is not to find an ethical discourse to regulate your contexts according to some kind of derived principles or rules (303). Instead, the whole orientation toward this structure has to change: "Learning III (*i.e.* learning *about* Learning II) may lead either to an increase in Learning II or to a limitation or perhaps a reduction of that phenomenon. Certainly it must lead to a greater flexibility in the premises acquired by the process of Learning II—a *freedom* from their bondage" (304). In L3 one does not entirely avoid building and acting according to contexts, since such things are undoubtedly useful (maybe new shoes would help me finally get into running; it's probably best if I do not do a headstand in the airport), just as L2 does not eradicate L1. But in learning about this process of forming contexts we can start to see these linkages, as Augustine called them, and open them up for examination and transformation.

Crucially, identity is essential to the whole operation. L3 requires the release of your sense of limited self:

> But any freedom from the bondage of habit must also denote a profound redefinition of the self. If I stop at the level of Learning II, "I" am the aggregate of those characteristics which I call my "character." "I" am my habits of acting in context and shaping and perceiving the contexts in which I act. Selfhood is a product or aggregate of Learning II. To the degree that a man achieves Learning III, and learns to perceive and act in terms of the contexts of contexts, his "self" will take on a sort of irrelevance. The concept of "self" will no longer function as a nodal argument in the punctuation of experience. (304)

The entirety of L3 hinges on the possibility of bringing into view the artificiality of one's self-identity. To learn on this level requires that this identity be open for question and moderation, or even rendered irrelevant. This relationship between L3 and identity becomes particularly important for our mode of speaking of ideologies and prejudice, particularly since the currently dominant modes of discourses tend to reinforce identity structures according to particular categories. From the perspective of these learning levels it ought to be clear that the widespread biopolitical regulation of identity is an efficient means for political structures to direct power.[36] As long as we continue to function primarily in terms of identity, structures of power will be able to tune those identities in order to reorganize economic and power flows in ways that can be partial and parasitic. However much we attempt to change the organization of those contexts and identities, people will always find themselves falling into harmful modes of categorization and holding to their acquired preferences (L2). Without some attempt at L3, prejudices will always appear as indomitable structures that we fail to locate because we cannot see them within our own identity structures.

Bateson notes that L3 is ultimately difficult to achieve and potentially dangerous, since when it is only performed partially it can lead to an instability. This makes sense even in our basic example of racial prejudice: one notices the prejudicial contexts within oneself and, unless performed thoroughly through the release of identity, one immediately places them within the context of L2: "Oh no, I'm (not) a racist!" Such partial means of noticing one's contextual and identification structures surely contribute to the cycles of guilt, frustration, anger, annoyance, and avoidance that we perpetually move between within discussions of identity politics. In many ways, this accounts for the "I am sinful and corrupt" tone of much of religion as we already see in Augustine's *Confessions*. The sense of self-corruption is the birthing pain of a consciousness beginning to arrive at the capacity for L3. In turn, social structures, such as organized religions, can take advantage of this beginning by tying such observations back into a system of self-identification that prevents full L3 from occurring.

However, if fully consummated, L3 induces a noticeably scalar experience that Bateson describes in contemplative terms:

> The resolution of contraries reveals a world in which personal identity merges into all the processes of relationship in some vast ecology or aesthetics of cosmic interaction. That any of these can survive seems almost miraculous, but some are perhaps saved from being swept away on oceanic feeling by their ability to focus in on the minutiae of life. Every detail of the universe is seen as proposing a view of the whole. (306)

Here we find another articulation of the *tat tvam asi* relationship by which we realize ourselves as a localization of the Cosmos, but now clarifying how this new state of consciousness transforms habits, prejudices, and beliefs.

This structure of learning helps us understand Michel Foucault's observation that "know thyself" is given priority over the "care of the self" in Plato and Platonic traditions.[37] This priority is also mirrored in Advaita Vedanta and many forms of Buddhism precisely because the capacity to take care of one's "self" hinges on the deconstruction of the ego. Otherwise, "caring for oneself" will always be based on a conception of oneself, in the secondary self-narration of the DMN, and trapped in L2. This egoic caring of oneself will always be out of step with the dynamic and shifting realities around (ecologically) and within (in cells, atoms, and flows of energy) "you" as this ego attempts to act according to the narrative simplification built out of human-based experience. There is thus a disparity between the "self" implied by the two phrases: "care of the self" risks keeping intact what one thinks one is if in such care practices the "self" becomes the discernible body, the socially affirmed identity, the politically recognized subject, and so on (the options of identification posed in chapter 8). "Know thyself," on the other hand suggests that, in conjunction with any taking-care, one must inquire into this sense of self itself.

SURGICAL TECHNOLOGIES FOR COGNITIVE-EXPERIENTIAL IDENTITY EXTRACTION

We can now clarify how contemplative techniques might and might not be considered technologies of the self, which, in Foucault's description, "permit individuals to effect by their own means or with the help of others a certain number of operations on their own bodies and souls, thoughts, conduct, and way of being, so as to transform themselves in order to attain a certain state of happiness, purity, wisdom, perfection, or immortality."[38] The problem, again, lies in what one considers one's "own means," "their own bodies," and "transform themselves." What is this "self" on which these technologies are operating? Clearly, many such technologies tend toward self-help rhetoric and the neoliberal discourses of subjectivity that Foucault took such pains to diagnose. To the contrary, technologies of the self are better designated contemplative technologies, since they are about a direction of attention rather than the changing of an entity called "self." To target this notion of "self," such technologies must contain a reflective activity encountering what one thinks one is and what cognitive content is manifest within experience.

In such reflection, a split becomes apparent: when do we consider contemplative practices as a mode of improvement versus an inquiry into the nature of the self? The former requires a standard—some sense of a good and a bad—and therefore a judgment built out of a model of the world. Whether such standards are built from social mores, personal expectations, or scientific data, the situation is displaced from the scalar configuration from which experience emerges. Such standards risk imposing on this experiential configuration descriptions derived from out of its scale. Remember, experience has already become overwritten even as it is already delimited in certain ways. Thus, you think you are the body, with a certain name, attributes, tendencies, needs, obligations, memories, and abilities. Yet these linkages were already built on a reality that exceeds this parsing; within your experience are already cells and ecologies and atoms and stardust. Can we sit in this tuning that both acknowledges this experience here while also transforming it, but not merely as another imposition of an external narrative (an L2, DMN activity)? The great risk is that these new scalar objects become another imposition and overwriting as we trace DNA ties, ecological impacts, toxic molecules, heat deaths, hormone traces, risk probabilities, autoimmune effects, pandemic statistics, and evolutionary factors. We thus overload our "selves," attaching more and more of this scalar configuration to this supposition that we are in control, that we must fix it, that we must know what to do next, how to improve our lives, the planet, our society, in order to adequately connect up to and account for these scalar objects we are continually uncovering. To the contrary, if we recognize that these connections and levels already are us, then we can calm down and examine the problem a bit more carefully. Where is the disorientation? How do we perform the reconfiguration?

The primary process must therefore be an examination of our sense of self so that we can understand what we have attached our notions, priorities, fears, and obligations to. This revision of contemplative inquiry is exemplified in Ramana Maharshi's distilling of meditation to self-inquiry, *ātma vichāra.*[39] This basic practice is to simply pose the simple question "who am I?" Of course, this is nearly the same as "know thyself" but engaged as a question rather than a command. Most importantly, this question is not conceptual. That is, one is not to produce any *content* or narrative about what one is. Rather, it is a mode of looking, a directing the attention back onto itself. *Ātma vichāra* calls you to observe within consciousness when content arises and look—"to whom does this arise?" The answer, undoubtedly, will be "me." But if you look again—who is this "me"?—one finds a blank, a nonanswer, an unknowing. The reason for this should be clear: this "self" is already an idea, a division, and an interpretation of the structure of experience. Looking for the self is to return the structure of thought to the continual flow of existence.

A similar version of self-inquiry is found in Plato's *Alcibiades,* structured as an inquiry into the division between tool and user, beginning with tools, then the hand, then the body, at each juncture noting the separation between that which is being used and the user. The problem is that Plato leaves us with the ambiguous term "soul" (*psukhē*), noting that "either man is nothing at all, or if something, he turns out to be nothing else than soul" (130c). The difficulty here, as in all of Plato, is the extent to which the dialogues appear to be satisfied with anything short of a full emptying out of the ego function (man being, in an important sense, "nothing at all"). Thus, Foucault suggests in his reading of this passage that "it is not at all the soul-substance he discovers, but rather the soul-subject."[40] In Vedanta, Shankara's Ātmashatakam (Song of the

Self) is nearly the same experiment, but it more carefully targets content that one might identify oneself with.[41] More importantly, it does so as a negation—I am not the mind, the body, the intellect, and so on. This negation serves as a movement away from this content or narrative. The positive statement, repeated at the end of each stanza, is then a direct identification with substance: *citānandarupah śivo'ham*—I am the form (rupa) of consciousness *(cit)* felt as bliss *(ānanda)*; I am Śiva (the Whole). This more clearly positions the "doer" of the *Alcibiades* experiment in the scalar configuration that refuses any subject less than the Whole itself.

The American philosopher and mystic Franklin Merrell-Wolff further clarifies the essential distinction as two kinds of reflection. He distinguishes this "finding content" as introspection: "Introspection remains far too objective (oriented to objects) to lead to the discovery of the self, with the result that many psychologists never find the true self. They do find something that they call the 'subject,' but they . . . have found only a subtle object."[42] This is the operation we identified earlier as self-help, the identification of some content within cognition or experience to identify as oneself to be cared for. Introspection of this sort needs to be distinguished precisely because it indulges the egoic, DMN narration operating at L2. In contrast, Merrell-Wolff provides the term *introception* to distinguish the "process whereby consciousness turns upon itself and moves towards its source" (104). Introception and introspection are two different operations with very different results. Literally, what they are looking at is different:

> The difference may be suggested by the figure of a lamp with a light within it. The subject [found in introspection] is only the lamp, while the self . . . is the light itself. Introspection, in the usual sense, can go no further than the lamp, insofar as it is the light that illumines and makes possible the subtle observation of introspection. . . . In introspection consciousness has not really turned upon itself, but merely establishes a kind of short circuit in the psyche. (103)

This is the "short circuit" already noted by Bateson as the manipulating L2 to masquerade as L3. In introception, by contrast, one directs one's attention to the flow of experience itself, "abstracting the subjective moment" apart from any content.[43] Merrell-Wolff's basic experiment is to look at an object and then, without withdrawing the perception of the object, draw the attention away from the object to the awareness itself (145–46).

The brain seems to have feedback mechanisms for identifying when self-inquiry or introception has occurred. Thus there are a surprising number of resonances between terms that describe an element of this scalar operation but then come to mean bliss or realization. An epiphany is the unveiling or manifestation of a larger entity—*epi* meaning "above" and the Greek *phania* meaning "to appear." Similarly, ecstasy is felt when one is literally "beside oneself" in the moment of egoic dislocation. In Sanskrit the term *samādhi* is likewise translated as "bliss" but literally means "to bind together." The common element here—a feeling of overwhelming bliss—is clarified by the Sanskrit term *ānanda*. The Ribhu Gita suggests that *ānanda* is a kind of feedback via the mantra *ānandam paramam mānam idam dṛśyam na kiñcana* (bliss is the ultimate measure, this which is visible is nothing at all).[44] Within the reorientation to the non-egoic consciousness, bliss is a natural feedback of the brain resituating itself in accordance with its nondivided reality. Love or the Greek *agapē* serves the same function within Christian contemplative practices, as can be seen in the *Cloud of Unknowing*: "Then perhaps he may touch you with a ray of his divine light which will pierce the *cloud of unknowing* between you and him. He will let you glimpse something of the ineffable secrets of his divine wisdom and your affection will seem on fire with his love."[45] Like *ānanda*, the love itself both serves as a sign of the transformation and further develops it. In short, it is what it feels like to retrain oneself in the non-dual, non-egoic consciousness: "The work of love will eventually heal them totally."[46] The final result is best described as a *metanoia*, meaning a change or alteration of mind. This term, which is cited by both Aldous Huxley and Edgar Mitchell to describe their experiences, is usually translated as "repentance" or "conversion," but is more properly the transformation induced by the persistent reorientation that brings one beyond the limited mind constructs of egoic narration.[47]

From this perspective, we can understand Augustine's description of his conversion, which is written in noticeably scalar terms:

> Being thus admonished to return to myself . . . I entered into my inmost being. . . . I entered there, and by my soul's eye. . . . I saw above that same eye of my soul, above my mind, an unchangeable light. . . . [This light was not] above my mind, as oil is above water, or sky above earth. It was

above my mind, because it made me, and I was beneath it, because I was made by it. He who knows the truth, knows that light, and he who knows it knows eternity. Love knows it. (132)

Augustine's conversion here is a *metanoia* in which he experienced that-which-is-behind the part of consciousness incorrectly identified as "his." This moment of reflection, introception, and self-inquiry is then reinforced by the feedback of love and awe, which leads him on to the final subsumption of his will to the unfolding of the One-Whole-All.

While self-inquiry is the basic operation and necessary element, fully retraining the brain might require additional techniques and adjustments that handle the particular contents of identification as they arise. Inevitably, introception is difficult because it operates on what arises as "me": the content built most strongly into the structure of the brain. Such content must be persistently extracted by any means until the cognitive apparatus has been completely reoriented. However, the techniques needed are particular to any given cognitive-experiential structure; there is no one universally applicable technique or practice for performing the extraction. Thus, Merrell-Wolff notes that, while many methods and techniques provided by contemplative traditions will likely prove helpful, "the successful method always had to be in some measure an original creation."[48] Similarly, the *Cloud of Unknowing* notes that "techniques for controlling your thoughts are better learned from God through experience than from any man in this life."[49] Since the loop is within the cognitive structures manifested here, where the egoic construct has taken hold and divided itself off, there is nowhere else to look to refine, experiment with, and perform this extraction.

I, ENDING IN THIS ECOLOGICAL RESIGNATION

In Evelyn Underhill's foundational account of the mystic way, she speaks of a kind of initial awakening that must then be developed and integrated.[50] With scale we repeatedly find ourselves brought into contact with powers smaller and larger than our human limits, entities that awaken us to a world beyond but already within the limited lifeworld present in everyday experience. To some degree, then, we already find ourselves halfway to such an awakening; what is needed is not a new realization but a turning toward and a recognition—it's already here, in your face, in your cells: you are not contained; all of this is connected. You have felt some glimpse of it already, but this insight has been redirected in a thousand ways (to the apparatus of the state, which claims this status of higher power; to the reification of a monetary system whose abstract structure of value seems to provide a sense of order; to the claim of a religious structure that will fit this realization into a set of prescribed beliefs; to the intellectual posturing of critical analysis without self-transformation).

Writing at the beginning of the great acceleration called the Anthropocene, Søren Kierkegaard provides a glimpse of this situation in his insistence that we not move on so quickly past the basic points. In his dwelling with the paradox of Abraham's near-sacrifice of Isaac, he describes this most difficult attitude that permits you to move forward even when YHWH, the being that is, forces you to face a situation beyond the rational. Situated between the this-scale multiplicity (the finite) and the scale of the One-Whole-All (the infinite), "with infinite resignation he has drained the cup of life's profound sadness, he knows the bliss of the infinite, he senses the pain of renouncing everything, the dearest things he possesses in the world, and yet finiteness tastes to him just as good as to one who never knew anything higher."[51] Is this attitude possible, as we struggle to face this vast Cosmos and find a way to love whatever is, however it unfolds itself?

Ecological crisis—not just global warming but all of the effects of our ongoing capitalist expansion and general unawareness of our collective accumulations—is a kind of collective Abraham moment. We are now within the three days where we walk to the mountain with our most prized Isaac—the enormous technological-economic-geopolitical edifice we find ourselves within—to be sacrificed on the altar of our own planet. In contemplating this situation, we find ourselves, like Kierkegaard, stuck in the absurdity, unable to act as Abraham, who still resolutely lives and loves this present even in the impending task. Since it is not any one of us humans who will learn to make the adjustment to something called "sustainability," the collective tide of these human-bound systems will determine whether the knife will be withheld. Facing this scalar loop, we tremble and ask: will it be so if we remain despairing at the approach? Will we remain paralyzed here by our own angst? Or will the adjustment only be made within a new kind of faith, one that peers into the infinite variety of existence and does not reel in horror or jump forth in conquest or get lost in the reconfiguration but learns to dwell with this cosmic vision?

Such faith is a dwelling in the involution. Without it, I risk intellectually knowing that I belong to this larger structure without losing myself in it. How can we tune our ability to respond to this world in this skewed stance? Logically, I can speak of how we are all part of the same planet and therefore have some kind of obligation to it. Yet in this very assertion I can fail to see the power in the view itself and fail to allow the scalar relation to rework how I live and view my existence. We are speaking here not only of those who do not comprehend their role in a larger environment but also those who proclaim to care about the environment and invoke such concern socially, habitually, insistently, without taking on the possibility of scaling all the way, placing oneself on the line, and experiencing the self itself as ecological.

Today, this cosmic inoculation—being exposed to this scalar configuration just enough to either ignore it, trivialize it, or simply proclaim it—is the major obstacle to our ability to see and feel our place in the Cosmos. We do not simply need to see more or better, but in a qualitatively different fashion. We need to learn to spend time contemplating these scalar relations, but doing so requires that we each take on the difficult task of looking at and facing the fundamentally false sense of egoic identification that lies at the core of our interpretation of the world. Once we take seriously that these relationships have to be, in some way, performed on oneself—that a mere glimpse is not enough, even if the experience manifests as a cosmic glimpse—then perhaps we can open ourselves to experimenting with the scalar movement. To do so, the personal reference point must for one moment, and then another, and then another, be allowed to disappear, to permit the scalar situation to come into view and reorient the structure of cognition around this new information. Only then will these *Homo sapiens* align themselves to this complex reality that they are already a part of. Only then will we find our collective systems drifting—colossally, meteorologically, Whitman would say—toward a world, as Taylor so hopefully describes, filled with beautiful, peaceful, compassionate, loving people who, rather than trapping themselves in a limited and failing fragmented mentality, recognize that they operate as and within the Whole.

Acknowledgments

The retrospective appraisal of any pathway through this-scale experience always appears both improbable and fortuitous. To the many people (not to mention plants, animals, bacteria, and energy gradients) who went out of their way (literally altering their pattern of existence) to generate the context for this textual unfolding, I can only respond with gratitude and awe.

This book is dedicated to Richard Doyle, who provided a baffling amount of energy, wisdom, confidence, and always unfolding insight not just for this project but for the whole locality called "me." The project would not exist without his generous attention, his unwillingness to let me compromise the integrity of the project, and his insistence that I go more slowly. His influence can be felt on every page.

As an even more lost and youthful wanderer, I also owe much to Pat Gehkre, who took me under his wing and showed me that this was possible, and to Kevin Eliot, Shevaun Watson, Laura Walls, Jill Frank, Alfred Nordmann, Erik Doxtader, and Ann Johnson for putting up with my enthusiasm while introducing me to the life of the mind. Along the way other interlocutors and teachers appeared, especially Jeffrey Nealon, Mark Morrison, John Jasso, Susan Squier, Jack Selzer, Jeremy Engels, Jason Maxwell, Robert Birdwell, Curt Tigges, Michelle Huang, Ralph Cintron, Zach Horton, Derek Woods, Peter C. van Wyck, Anushka Peres, Kenny Easwaren, Keith Lloyd, and, in different capacities, Gary Weber and Erica Kaufman. Andrew Pilsch has been a particularly essential friend, colleague, and supporter. Special thanks to those in the sciences who provided thoughts relevant to this work, including Mark Shriver, Katie Barlow, Joseph Johnson, David Saint John, Dan Sykes, Sadhat Walusimbi, Tianyanxin Sun, and Joy Pate. A special thanks to my students, especially my undergraduate assistants Mutafa Al-Nomani and Hannah Mailhos, for helping with the manuscript, as well as Anna Rogers, Jackson Bell, Hannah Green, and Jackson Greer, who took on research projects related to my own. A great deal of insight was developed with the graduate students in my "Rhetoric after Information" course: Tyler Champine, Kristian Cantens, Jessie Cortez, Savannah Foreman, and Erika Rodriquez. Thank you to Cary Wolfe for assisting at a crucial juncture in the publication process. This book would have not made it to the current form without the support of my editor, Doug Armato, his assistant Zen Miller, and the two rounds of insightful reviews from Phillip Thurtle as well as an anonymous first-round reviewer.

I want to acknowledge my department at Texas A&M, which has generously supported my work and been not only willing to entertain but excited by my disciplinary wandering. Parts of the research for this project were supported by the Penn State Center for Humanities and Information. Although no archival materials are cited here, I want to acknowledge Philip Guddemi, the Bateson Idea Group, and the UC Santa Cruz Special Collections for an enlightening visit to the Gregory Bateson archive. Thanks to Ron Leonard and Dorothy Leonard for making available the work of Franklin Merrell-Wolff, whose conceptual clarity was essential to the shaping of this book. Thank you to those involved in Zebrapedia and the transcription of Philip K. Dick's *Exegesis,* including Rich Doyle, Max Spiegel, and the Estate of Philip K. Dick, who made the work available. Thanks to Anil Mundra for your patience as a Sanskrit instructor and for taking the time to read a draft. I should also acknowledge the Perseus Digital Library, hosted by Tufts University, which was an essential tool for carefully examining Plato's texts for someone whose education has not yet afforded the opportunity for lessons in Greek.

Finally, thank you to Sara DiCaglio for your tremendous support and patience, academic and otherwise, for putting up with years of musings about scale, and for enduring the Texas heat so that this can continue to happen. And to my children, Milo and Penelope: thank you for making the past few years as fun as they were intellectually stimulating. May we all learn to feel as much wonder at the scales of existence as you both expressed in first experiencing the *Powers of Ten.*

Notes

INTRODUCTION

1. Levin, "The Problem of Pattern and Scale," 1944. Compare Herod's statement to the same effect about geography in *Scale* (xi).
2. Levin, "The Problem of Pattern and Scale," 1945.
3. R. V. O'Neill and King, "Homage to St. Michael," 5–6.
4. See Jazairy, "Towards a Plastic Conception of Scale," 1; Tong "Ecology without Scale" 198; Herod and Wright, *Geographies of Power,* 5.
5. Herod and Wright, *Geographies of Power,* 5.
6. Tong, "Ecology without Scale," 198.
7. Nigel Thrift quoted in Marston, Jones, and Woodward, "Human Geography without Scale," 416.
8. Milburn, *Mondo Nano* and *Nanovision;* Hanson, *Haptic Visions.*
9. Myers, *Rendering Life Molecular;* Helmreich, *Alien Ocean;* Hird, *Origins of Sociable Life.*
10. Hayles's *How We Became Posthuman* has had the most enduring impact of her work. Clarke has published widely on cybernetics and systems theory often directly connected up with questions of ecology and Gaia (e.g. *Earth, Life, System*). Wolfe has discussed systems theory extensively, particularly in *Critical Environments.* Doyle's work has crossed multiple scales; *On Beyond Living* and *Wetwares* deal primarily with the encounter between informatic and biological scales, while *Darwin's Pharmacy* handles scale far more explicitly, connecting the discourse of evolution with ecology and psychedelics in a necessarily cross-scalar relation.
11. Heise, *Sense of Place;* Messeri, *Placing Outer Space;* Morton, *The Ecological Thought* and *Hyperobjects.*
12. Scholarly interest in cultural depictions of scale was greatly sparked by Timothy Clark's essay "Scale: Derangements of Scale" and Derek Woods's "Scale Critique for the Anthropocene." Two edited collections have appeared (Clarke and Wittenberg's *Scale in Literature and Culture* and Carr

and Lampert's *Scale*) as well as a special issue of *Modernism/Modernity* (see Davis and Hensley, "Scale and Form") partially inspired by Tung's "Baddest Modernisms."

13. See especially the introductions to Clarke and Wittenberg's *Scale in Literature and Culture* and Carr and Lampert's *Scale*. Both of these largely import geography's notion of scale (see chapter 9, endnote 24, and chapter 11, endnote 8). Most importantly, this work does not distinguish between what we will call Gulliver's scaling and cartographic scaling from scale itself, largely because the disciplinary investments here are focused on representations and narratives of scale.
14. Pattee's 1973 edited collection, *Hierarchy Theory*, provided a starting point for this conversation. These hierarchy theories have precedent in the organismic philosophies of the earlier twentieth century, e.g., in Bertalanffy's *Problems of Life* (chapter 2).
15. See Salthe and Fuhrman, "The Cosmic Bellows," and Ahl and Allen, *Hierarchy Theory*. For a more recent summary of concepts of scale see Wu and Li, "Concepts of Scale and Scaling."
16. Primack and Abrams, *View from the Center*, 294.
17. For an overview of the struggles to define mysticism see Ron Leonard, *The Transcendental Philosophy of Franklin Merrell-Wolff*, 78–93. This definition of mysticism is unusual and may not seem immediately apparent to those who study mysticism. The reason is partially because this definition requires this theory of scale to see all of the overlaps between mysticism and scale. Providing this theory of scale will, I hope, clarify some essential things about mystical experience even as we focus here primarily on scale. What follows will show how certain experiences that seem to contradict or seem exceptional might be part of the same set of transformations in consciousness. For example, contrary to Jess Beck Hollenback's claim that "mystical experiences have no essential connection to states of 'union' or 'oneness' with the Absolute" (*Mysticism*, 132), in chapter 7 we will see how such oneness fits with a whole configuration toward reality that includes many of the aspects that Hollenback emphasizes in her study of mysticism.
18. Versluis, *Platonic Mysticism*, 7–8.

1. DISTANCE AND RESOLUTION

1. The manner in which the eye makes units was fundamental to Gestalt psychology, and becomes essential to both the science of vision and visual technologies, which take advantage of these aspects to produce visual

representations. Thus it is no surprise that these aspects of unification of seemingly disparate parts becomes essential to a science of vision, e.g., in the early computational theory of vision in David Marr's *Vision*. Indeed, Marr's "underlying physical assumptions" include a specification of scale by the visual system (45).

2. While scientists separate out many more factors in delineating the object of study and what is being viewed, the definition provided here fits generally with the way the term "resolution" is used. For example, one definition used by geographers is "the smallest distinguishable parts in an object or sequence" (Lam and Quattrochi, "On the Issues of Scale," 89). A similar definition has been used in metrology: "the smallest difference δ between indications that can meaningfully be distinguished" (Lira, "Resolution Revisited," L14). In microscopy, the concept of resolution is defined likewise as "the ability to differentiate between two small, closely spaced objects" (Ryan et al., "Introduction," 2).
3. This remixes Gregory Bateson's definition of *information* as "a difference that makes a difference." The point becomes one of the foundations for the scalar reconfiguration of objects in chapter 7.
4. For reasons that will become clear, it is essential that we reserve the term "Whole" here only for this largest-scale perspective This use of "Whole" runs us up against a history of uses of "whole" in science that differs in essential ways from this definition (see Harrington, *Reenchanted Science*). Throughout this work I will mark this specialized notion of the Whole, All, and One with the capital in order to distinguish them from the more limited ways these words can be used.
5. See *Bhagavad Gita,* 12:3–4. Definition from Monier Williams.
6. I am using "negation" primarily in the sense implied by traditions of negative theology, particularly Pseudo-Dionysius: the act of voiding the essentiality granted to a category. Although the term is far more loaded, I suspect that this use fits to some degree with Hegel's use of the term, since we are, in some way, coming to a "negation of the negation."
7. Some of the theories listed here more or less consider this singularity as tentative or functional operations. However, the problem, which we will face extensively in chapter 7, is that *we are perpetually forgetting* this secondary character of individuation and treating things as having an "in themselves."
8. This idea has also been noted by Woods in "Epistemic Things," 78–79.
9. See especially Haldane, "On Being the Right Size"; Bonner, *Why Size Matters*; West, *Scale*.

10. As we will discuss in chapter 11, Tsing critiques these in "On Nonscalability."

2. MEASUREMENT AND PERSPECTIVE

1. Plotinus (*Enneads*, II.8.1) notes that we need to be able to account for distances and parts to accurately observe the magnitude of a distant object. See also Plato's discussion in *Protagoras* 356c-e.
2. Astronomers use this basic principle in combination with more complicated techniques to determine the size and distance of astronomical objects. These techniques are called the "cosmic distance ladder." See Webb, *Measuring the Universe.*
3. Brian Rotman has an excellent discussion of perspectival zeroing out while discussing the "vanishing point" in art. The "vanishing point" permits the representation of a "definite location within a real physical scene" while remaining a "visual zero" (*Signifying Nothing*, 19). However, Rotman's interest is how this vanishing point offers the spectator the possibility of objectifying himself, the means of perceiving himself, from the outside, as a "unitary seeing subject" (19). We are offering it here in a different sense—as the starting point for accounting for a perspective.
4. The language here is drawn in response to Haraway's "Situated Knowledges," discussed further in chapter 8.
5. I would like to thank the presenters at the 2017 English Institute for examples and discussion that led to this point, particularly those by Mary Ann Doane, David Wittenberg, Ursula Heise, and John Durham Peters. Some of these essays have been published as Macpherson and Martin, "Essays from the English Institute 2017."
6. See Myers, *Rendering Life Molecular*.

3. SCOPE AND ACCUMULATION

1. If we already assume scale, the first (relations moving across a boundary) can be produced by going to a larger scale and the second (more already in a relation than is originally discerned) can be produced by going to a smaller scale. As an example of these together, one article tying together geomorphology and ecosystem ecology notes that "because most ecosystem processes are scale dependent, the choice of boundaries is of profound importance to the conceptualization of an ecosystem and to the scope and validity of questions being asked within that ecosystem. Indeed, any conceptualization of an ecosystem requires constraining the spatial and temporal scales of analysis" (Post et al., "The Problem of Boundaries," 111).

The turn to scale here arises from the need to delimit and account for the contingency and porousness of the attempt to define an ecosystem even according to geomorphology.

2. Haldane, "On Being the Right Size."
3. From the perspective of Gaia theory, the Earth is already integrated into organism-like feedback loops (see chapter 9). However, Vernadsky and Teilhard de Chardin push further, considering man part of an emerging noösphere, which suggests more integration is forthcoming. See Vernadsky, "The Biosphere and the Noösphere"; Teilhard de Chardin, *The Phenomenon of Man;* Doyle, *Darwin's Pharmacy,* 175.
4. Set theory usually bans the set of all sets because it creates a logical paradox. As we will discuss in chapter 7, we can say that there are two kinds of abstraction that must be distinguished if we are going to talk about the "highest order of abstraction": abstraction in signs and abstraction about signs. Cantor's "set of all sets" is an abstraction in signs. The highest order of abstraction meant here is not merely a sign ("set of all sets") but a sign about signs: it is a glimpse of the process of abstracting itself whereby one designates sets. It is a sign that indicates something about set-making, specifically the openness to anything that might be found in distinguishing sets. Thus, "set of all sets" *as a term* is of a different logical type than any set. We only think that this continues to produce paradox if we ascribe too much concreteness to the phrase "set of all sets" (Whitehead calls this problem "misplaced concreteness").
5. Hilbert's Hotel was formulated in 1924 but was popularized by Gamow in *One, Two, Three . . . Infinity,* 17–18.

4. TO THE BOTTOM

1. The Planck length is mathematically derived from three known constants (the speed of light, the gravitational constant, and Planck's constant). It is theoretically the smallest scale at which gravity would have an effect or where the known aspects of the universe would be relevant. This length is incredibly small: 1.616255×10^{-35} meters.
2. There are many hints of this formulation in physics, for example in Bohm and Hiley's ontological interpretation of quantum physics. Chapter 14 of *The Undivided Universe* hints at scale as relevant for the resolution of some paradoxes of quantum theory. For example, this thought experiment provides a way of conceptualizing the "jumps" that makes Bohm's ontological interpretation possible as opposed to Bohr's epistemological one, allowing one to speak of a determinable state of a quantum even if it is not

predictable or controllable (see Bohm and Hiley, *The Undivided Universe,* 25; see also chapter 7 here).

3. The cellular automata of Wolfram's *A New Kind of Science* and the chemical creation myth of Kauffman (see *At Home in the Universe*) provide two other ways to think of this aggregation as a kind of patterning made possible by simple rules of interaction on a scale below (see chapter 9).
4. This point suggests a scalar aspect of the self-assembly of components, e.g., in Foerster: "The form of the elementary building blocks influences the possible forms of self-organization" (*Beginning,* 126). Similarly, Wolfram presents the possibility that "particles are analogs of the localized structure" within the kind of patterns he identifies in *A New Kind of Science*: "and if this is so, then it means that at the lowest level, the rules for the universe need make no reference to particular particles" (525).
5. For example, Bertalanffy: "Here, Newtonian mechanics and classical physics, as based upon these visualizable categories, are perfectly satisfactory. They break down, however, if we enter universes to which the human organism is not adapted. This is the case, on the one hand, in atomic dimensions, and in the cosmic dimensions on the other" (*General Systems Theory,* 242). Bohm and Hiley make a similar point in their discussion of the difference between classical and quantum world (*The Undivided Universe,* 347).
6. Accounts of the genesis of quantum physics often note that statistical methods were essential to making the transition to the quantum (e.g., Stone, *Einstein and the Quantum,* 40–41). Prigogine and Stenger's *Order out of Chaos* begins with observations about probability to this effect: "Obviously, in addition to deterministic processes, there must be an element of probability involved in some basic processes" (xxvii).

5. FROM THE TOP

1. There are two puzzling aspects of this chart that I am unable to provide any clarification for. First, why do the gaps between the scale domains vary to some degree? Does this indicate some error in how I have separated them here, or is this just the way scalar aggregation/differentiation functions? Second, why is there is so much space still left between the quarks and the Planck length? This indicates that there is plenty of room for the definition of more quanta, whether they be strings or some object yet to be designated—unless there is some reason that we find a much wider ranging scale domain at this lower resolution.

2. See Mandelbrot, *The Fractal Geometry of Nature*, 25–33.
3. This point has been made at length by Primack and Abrams in chapter 6 of *View from the Center.*

6. IN THE SCALAR SIMULATION

1. The term "zoom" is derived from the sound made by a fast-moving object. In one sense, then, it is an appropriate metaphor, since the orders of magnitude created by scale mean that we are changing perspective with greater and greater speed (of the sort performed in the first experiential origin). But, in another sense, it is inappropriate, since this is not a movement at all, but a change of resolution.
2. We will address these critiques more in chapter 11.
3. If we ran this difference through Foucault's analysis of late-eighteenth-century medicine in *The Birth of the Clinic*, we can find a curious question about this transition point Foucault is identifying. Foucault's signature line here expresses it quite well: "If you dwell on the surface, nothing is clear, but open up a few corpses and it starts to become clear" (146). But if going beneath the surface is not just a going into the body but smaller in scale, as it becomes in the subsequent decades following Bichat (especially in the development of cell theory and later with Pasteur), then we find an even more significant shift in medicine. We will discuss this in terms of cell theory in chapter 9.
4. Solé, *Phase Transitions*, 11.
5. Tomes's *The Gospel of Germs* provides an excellent historical account of our struggle to adjust to this scalar cause.
6. Wu and Li, "Concepts of Scale and Scaling," 3.
7. This was the kind of biology instantiated by D'Arcy Wentworth Thompson's *On Growth and Form*, first published in 1917, which bridges the mathematical and the developmental. This has developed into the field of evolutionary developmental biology (evo-devo), which considers the many different ways constraints and aggregation work in forming biological systems.
8. Bateson, *Mind and Nature*, 44, 41.
9. The notion of *Umwelt* comes from Uexküll's *Foray*.
10. Primack and Abrams, *View from the Center*, 172.
11. See, for example, the final chapters of West, *Scale*, and Kaufmann, *At Home in the Universe*.

7. IN-FORMATIONS OF THE WHOLE

1. Dick's *Exegesis* (4:106; 48) contains an account of his Valis experience. The story was first made available in *VALIS* and "How to Build a Universe." Following the practices of the published edition of the *Exegesis* (ed. by Pamela Jackson and Jonathan Letham), all citations to the *Exegesis* will be given as (folder number: page number). If the quote appears in the published volume of *The Exegesis* the pagination from that edition will follow a semicolon. The full documents can be found at zebrapedia.psu.edu.
2. Dick, *Exegesis,* 693n.
3. Elsewhere I have more extensively discussed how to read the *Exegesis* within the frame of scale and mysticism presented here. See "Simulations of Moksha."
4. Wolfe, *Critical Environments,* x.
5. The phrase "take the God out" is borrowed from Rotman's subtitle for *Ad Infinitum.*
6. On information theory in poststructuralism see Geoghegan, "From Information Theory to French Theory."
7. Merrell-Wolff, *Transformations in Consciousness,* 189.
8. Bateson's oft-cited definition can be found in the glossary to *Mind and Nature* (228), as well as in *Steps* (315).
9. In contrast, Dick references a different kind of "ethical balking" at the world, which he says is the necessary starting point for beginning to understand this configuration (*Exegesis,* 32:3; 271)
10. At 1.19 we distinguished differentiation and division to separate out notions of division as an ontological separating of parts of reality into new objects and differentiation as the epistemological selection of differences to designate a separate object. In this section, however, I will be using "division" to more adequately highlight the difficulty these epistemological questions pose for ontology—for what we consider separate in itself.
11. Heraclitus, Fragment 50 (118 in the Wheelwright translation). *Logos* is an important term in the history of Greek philosophy and a particularly important term for Heraclitus. It is variously translated as "to speak," "reason," or "account." For Heraclitus, Logos is the name for this larger structure of existence. I prefer this translation from the classicist Wheelwright; the point is much the same in Robinson's more literal translation, but he parenthetically weakens the statement of oneness: "Not after listening to me, but after listening to the account [logos], one does wisely in agreeing that all things are (in fact?) one (thing)" (37, brackets in the original).
12. Dick, *Ubik,* 226.

13. Fragment 8, Gallops, *Parmenides.*
14. Dick paraphrases Heraclitus as saying "we woke up from our dream and began to see reality, i.e. the plan or logos" (5:43; 140). This seems to be a liberal gloss of Fragment 1: "Although all things come to pass in accordance with this Logos, men seem to be quite without any experience of it. . . . Other men, on the contrary are as unaware of what they do when awake as they are when asleep" (Wheelwright translation). Dick's rendition may be derived from Hussey's description of Heraclitus as describing men as "sleepers in private worlds of their own" (*The Presocratics,* 38).
15. Dick, "The Indefatigable Frog," 221.
16. Schrodinger: "Which part, which aspect are *you?* what, objectively, differentiates it from the others? No, but inconceivable as it seems to ordinary reason, you—and all other conscious beings as such—are all in all. Hence this life of yours which you are living is not merely a piece of the entire existence, but is in a certain sense the whole; only this whole is not so constituted that it can be surveyed in one single glance. This, as we know, is what Brahmins express in that sacred, mystic formula which is yet really so simple and so clear: *Tat tvam asi,* this is you. Or again, in such words as 'I am in the east and in the west, I am below and above, I am this whole world'" (*My View of the World,* 21–22).
17. Schrodinger, vi–vii.
18. Bohm, *Wholeness and the Implicate Order,* 12, emphasis added. Cf. Plotinus: "In our realm all is part rising from part and nothing can be more than partial; but There each being is an eternal product of a whole and is at once a whole and an individual manifesting as part, but to the keen vision There, known for the whole it is" (*Enneads,* V.8.4 [489]).
19. Bohm, *Wholeness and the Implicate Order,* 17.
20. Bohm and Hiley make clear this role of interpretation when arguing for their theory of quantum physics, noting that it leads to the same mathematical and experimental results as others that have been proposed (*The Undivided Universe,* 5).
21. Hussey, *The Presocratics,* 92
22. From this point on, I capitalize *Wholeness* when it indicates this property of undivided existence at the scale of the One-Whole-All.
23. Hussey, 143.
24. References to Bucke appear in central locations in James, Huxley, and Underhill (James, *Varieties of Religious Experience,* 61, 294; Huxley, *Perennial Philosophy,* 63; Underhill, *Mysticism,* 253). Aurobindo uses the term "cosmic consciousness" in multiple writings (see, e.g., the chapter

titled "Cosmic Consciousness" in *The Synthesis of Yoga*), and Maharshi was asked about cosmic consciousness on his mountain near Tiruvannamalai (Maharshi, *Be as You Are,* 20). The Russian mystic P. D. Ouspensky addresses Bucke, as does the American Franklin Merrell-Wolff (Ouspensky, *Tertium Organon,* chapter 23; Merrell-Wolff, *Franklin Merrell-Wolff's Experience and Philosophy,* 12, 23, 69).

25. Bucke, *Cosmic Consciousness,* 17.
26. Bucke, 81.
27. Bucke, 81.
28. Lovejoy, *Great Chain of Being,* 13.
29. James, *Pragmatism,* 57.
30. James, 60. James writes of himself: "Whether my treatment of mystical states will shed more light or darkness, I do not know, for my own constitution shuts me out from their enjoyment almost entirely, and I can speak of them only at second hand" (*Varieties of Religious Experience,* 290).
31. Dick read Fritjof Capra and discusses QT in an interesting musing from April 1980 (*Exegesis,* 90:13; 705–7).
32. See Bricmont, *Quantum Sense and Nonsense,* 214–15.
33. Bohm and Hiley, *The Undivided Universe,* 13. Hereafter cited parenthetically in the text.
34. Bohmian mechanics, or the "de Broglie–Bohm theory," is an interpretation of quantum physics, stemming from and fitting the same mathematical and experimental results as other QT theories. Currently, Bohm's interpretation is marginal, as is discussed by Bricmont in *Quantum Sense and Nonsense,* despite being championed by widely respected physicists such as John Stewart Bell. The marginal nature of Bohmian mechanics only serves to highlight the ambiguity and the interpretive dilemma posed by scaling to the subatomic.
35. Bohm and Hiley: "In principle these links should extend to the whole universe, but for practical purposes their effects can be neglected on the large scale, so that in some suitable classical approximation we can use a simplified picture of the world" (*The Undivided Universe,* 252). Bricmont invokes this point to argue against hasty mystical conclusions, noting that "there are many things that happen 'in principle' in physics but that are negligible in practice" (*Quantum Sense and Nonsense,* 215). Similarly, Barad states: "It is not that we live our daily lives in a classical world, rather than a quantum one; the point is that we generally don't notice quantum effects because they are very small (too small to notice without special equipment)" (*Meeting the Universe Halfway,* 279).

36. Bohm, *Wholeness and the Implicate Order,* 22.
37. Bohm, 19.
38. See Barad, *Meeting the Universe Halfway,* 20. Hereafter cited parenthetically in the text.
39. Barad mentions in passing how "quantum physics is often positioned as the scientific path leading out of the West to the metaphysical Edenic garden of Eastern mysticism" (*Meeting the Universe Halfway,* 67).
40. For those familiar with the discussions of the two-slit experiments: Bohmian mechanics resolves the particle-wave duality in a brilliantly simple manner: the wave functions are described as a pattern that "guides" the particles that go through either slit so that they continue to produce a wave effect on the path of the particles even as those particles have particle-like attributes. For an accessible distillation of this solution see Bricmont, *Quantum Sense and Nonsense,* chapter 8. This description opens the possibility that some lower-scale wave behavior manifests as part of these larger-scale particle attributes in the manner already described in 4.4–12. Similarly, Bricmont's summary of the problem of observation is put in scalar terms: "Measurements do not really measure anything (except for detections of positions) but are interactions between a macroscopic system and a microscopic one" (161). Cf. Barad: "According to Bohr, apparatuses are macroscopic material arrangements through which particular concepts are given definition" (*Meeting the Universe Halfway,* 142).
41. See Bohm and Hiley's discussion of Bohr in *The Undivided Universe,* chapter 2.
42. Barad, *Meeting the Universe Halfway,* 144. On Bohr's avoidance of ontology see 125.
43. See the difference between difference and division at 1.19.
44. Morton, *Hyperobjects,* 15. Hereafter cited parenthetically in the text.
45. Morton, *The Ecological Thought,* 40.
46. Morton, *Hyperobjects,* 132.
47. Badiou, *Being and Event,* 25.
48. Lovejoy, *Great Chain of Being,* 51; Plato, *Timaeus,* 41c.
49. Lovejoy, *Great Chain of Being,* 51.
50. Lovejoy, 50.
51. Cf. the Isha Upanishad's reconciliation of the *sambhuti* (becoming) and *asambhutyam* (nonbecoming) in verse 12. See Aurobindo's discussion in *Isha Upanishad,* 23.
52. *Difference* is one of the primary terms of poststructuralist philosophy, foundational to Deleuze and Derrida and becoming the basis for

philosophies handling any kind of an Other in the political or social senses. I have focused here on Plato's formulation of difference, because it provides a surprisingly relevant account of the role of difference that is appropriate for this scalar account. While I'll leave it to others to trace out how this account of difference may differ from these philosophers', I can note that many emphases on difference, including Deleuze's, tend to invoke difference as generative of being. In one sense, we can agree that difference is generative of being in the sense of the Many, but this generation does not preclude the openness toward that being already implied in Deleuze and Guattari's plane of immanence. I'm arguing here that this openness is the predifferentiated One-Whole. This is indeed why Deleuze and Guattari have been accused by many of being mystics (e.g., by Hallward—see endnote 96 for chapter 9 below).

53. Ryle's "category mistake" is an application of the theory of logical types similar to Bateson's. He defines it using examples; e.g., "A foreigner visiting Oxford or Cambridge for the first time is shown a number of colleges, libraries, playing fields, museums, scientific departments and administrative offices. He then asks 'But where is the University?'" (*Concept of Mind,* 7).
54. Peirce, "On a New List of Categories."
55. See the discussion of Plato's *Protagoras* in the next chapter, where Plato invokes the need for a measure to track the inconsistencies in perception.
56. For example, 243b-c.
57. The longer Platonic tradition includes more adequate generalizations of this notion of illusion while retaining this relationship between the One and the Many. See, especially, Plotinus, IV.9 (418–22) and V.8.4 (489), e.g., "In our realm all is part rising from part and nothing can be more than partial; but There each being is an eternal product of a whole and is at once a whole and an individual manifesting as part, but to the keen vision There, known for the whole it is."
58. "Vajracchedikā Prajñāpāramitā," verse 32, trans. Paul Harrison. Cf. Sankara, *Direct Awareness of the Self,* verses 62–63: "Just as we think we see blue in the sky, mirage in the desert, or human figure in a post, so Brahman appears to us as the universe. But as a ghost appearing in empty space, a castle in the air, or a second moon in the sky are illusions, so is the universe" (82–83, trans. Douglas Fox).
59. Kant, *Critique of Pure Reason,* 88.
60. Merrell-Wolff took a graduate seminar on Kant's *Critique of Pure Reason* at Harvard. He often said that Kant did more than any other thinker to help

"prepare the way" for his *gnosis* (see D. Leonard, *Franklin Merrell-Wolff*, 16–17).

61. Merrell-Wolff, *Franklin Merrell-Wolff's Experience and Philosophy*, 5, 6.
62. Merrell-Wolff introduces consciousness-without-an-object on page 29 and elaborates on it, particularly in relation to Kant, in the latter half of his book. The discussion of the "in themselves" is addressed on 331–32 and 386–88. For a philosophical appraisal of this concept, see R. Leonard, *The Transcendental Philosophy of Franklin Merrell-Wolff*.
63. I first encountered this statement that consciousness is always "consciousness of" in Merleau-Ponty (*Phenomenology of Perception*, 43). It is echoed, albeit flipped onto a subject, in Morton's *Hyperobjects*: "What is called *consciousness* is an aesthetic effect: it is consciousness-for" (84). This distinction becomes fundamental in Gupta's account of *cit* and consciousness in Vedanta; she argues that this question of consciousness as strictly possessive is due to an assumption that consciousness must be intentional (*Cit: Consciousness*, 7, 101).
64. This invocation of consciousness is always controversial, but I invoke it here for the necessity of accounting for the potential for any part of "what is" to be described in terms of response-ability. Such invocations of "consciousness" as a general term are attempts to extend the scope of something we readily observe within observation—that being aware is an attribute of reality. It is interesting how readily we accept a generalization of other ways of positing reality, such as energy, resistance, weight, tangibility (i.e., matter), or even agency, but resist so strongly the same kind of description for the aspect of reality most immediately apparent within the inquirer. Regardless, a more nuanced version of this notion of consciousness cannot begin if we just simply dismiss it offhand. To that end, Gupta's *Cit* provides an essential exposition of this generalized notion of consciousness from an Advaitan perspective. See especially chapter 5.
65. Merrell-Wolff: "It is zero that symbolizes durable Reality, or Consciousness-without-an-object" (*Franklin Merrell-Wolff's Experience and Philosophy*, 386).
66. Nāgārjuna, *Nāgārjuna's Middle Way*, 13.7 (145). Citations are given as chapter and verse followed by the page number from the Siderits and Katsura translation in parentheses.
67. Nāgārjuna, 24.18 (177).
68. Nāgārjuna, 15.2 (155).

69. This notion of emptiness is not an exclusively Buddhist idea; it also exists in Sankara and in the negative theologies arising from Pseudo-Dionysius, e.g., in the *Cloud of Unknowing*'s notion of unknowing. This tradition already draws from Plato's *aporia* and statements of Plotinus—e.g., in his injunction that, if one would achieve the highest principle one must "cut away everything" (*Enneads*, V.3.17 [457]).
70. Nāgārjuna, *Nāgārjuna's Middle Way*, 13.9 (145).
71. See Plotinus, *Enneads*, V.1 (423–29).
72. Nāgārjuna, *Nāgārjuna's Middle Way*, 18.5 (197–98).
73. See Siderits and Katsura's commentary on pages 15 and 126 of Nāgārjuna, *Nāgārjuna's Middle Way*.
74. Merrell-Wolff, *Transformations in Consciousness*, 189.
75. Merrell-Wolff, 192.
76. Merrell-Wolff, *Franklin Merrell-Wolff's Experience and Philosophy*, 262.
77. Merrell-Wolff, *Transformations in Consciousness*, 195, emphasis added.
78. Merrell-Wolff, 190.
79. The other term of interest here is silence, *mauna* in Sanskrit, which has particularly important role in Advaita, e.g., in Sankara (*Direct Awareness of the Self*, 107) or in Maharshi (*Be as You Are*, 100–105). In this sense *mauna* is the silencing of the internal divisive process of thought or, in Merrell-Wolff's articulation, the investment in that thought.
80. Dick, *Ubik*, 793.
81. Dick, "The Indefatigable Frog," 228.
82. Dick, *VALIS*, 258.
83. Dick, 262.
84. Bateson, *Mind and Nature*, 99. Cf. Bateson, "Form, Substance, Difference," in *Steps to an Ecology of Mind*.
85. Bateson, *Mind and Nature*, 96–97.
86. Bohm, "Meaning and Information," 44. See also Bohm and Hiley, *The Undivided Universe*, 35–36.
87. Bohm, "Meaning and Information," 44.
88. Bohm, 55–57; Bateson, *Mind and Nature*, 6–10.
89. Bohm and Hiley: "While we are bringing out . . . the objective aspects of information, we do not intend to deny its importance in subjective human experience. However, we wish to point out that even in this domain, the notion of active information still applies" (*The Undivided Universe*, 36).
90. See Bohm and Hiley, *The Undivided Universe*, chapters 3 and 4, for a description of how this works within QT experiments.

91. Foerster uses the term in this way throughout *The Beginning of Heaven and Earth.*
92. Foerster, *Beginning,* 124.
93. This unit-making is already present in philosophies of information through aspects like state, threshold, or probability. Floridi's overview of definitions of information includes a number of potentially scalar conditions connected to this notion of probability, thresholds, and states: information is said to be a product of a system, a product of probability (either of a space distribution, as in Shannon and Weaver, or more broadly in the "probabilistic approach"), or about the state or exclusions provided within an aggregate, system, or person—i.e., some definable threshold (*The Philosophy of Information,* 31).
94. In Peircean semiotics this is called the interpretant.
95. Bateson, *Mind and Nature,* 68. Bateson's examples of overlaying perspectives include moiré phenomena, which was the basis of our experimental scalar climb at 4.7 (see *Mind and Nature,* 79–80).
96. Bateson, 70.
97. See also 1:46; 620—Dick will often use the overlay of two "times"—USA 1974 and Rome A.D. 45—as the guiding metaphor for this binocular vision. While this is stranger than our overlaying of scales, it fits with how you only need an awareness of two different points in space or time to produce the meta-abstraction.
98. Dick, *Divine Invasion,* 60, emphasis in original.
99. Merrell-Wolff, *Franklin Merrell-Wolff's Experience and Philosophy,* 65.
100. Bohm, *Wholeness and the Implicate Order,* xi.

8. I AM THE TRANSHUMAN COSMOS

1. Mitchell, *The Way of the Explorer,* 58.
2. Mitchell, 59.
3. In his foreword to Lyotard's *The Postmodern Condition,* Jameson states that there is a "characteristically French affirmation of the 'decentered subject' or the illusion of the coherent self or ego" (x). See also Cadava, Connor, and Nancy, *Who Comes after the Subject?,* for a compendium of responses to this question of the subject.
4. For a summary see Mansfield, *Subjectivity.*
5. Wolfe, *What Is Posthumanism,* xv.
6. Wolfe, xx.
7. "The belief that there is a polar opposition between Mind and Matter is the belief that they are terms of the same logical type" (Ryle, *Concept of*

Mind, 12). On logical typing see 3.6 or the discussion in chapter 7 (endnote 53).

8. Tanney's introduction to *The Concept of Mind* traces the ways Ryle has been interpreted, particularly in relation to materialisms.
9. Wilson, *Neural Geographies,* 1.
10. Gilbert, Sapp, and Tauber, "A Symbiotic View of Life," 326.
11. Wolfe, *What Is Posthumanism?,* 119.
12. Nagel, "What Is It Like to Be a Bat?," 435.
13. Dennett, *Consciousness Explained,* xi.
14. Chalmers first articulated this "hard problem of consciousness" in "Facing up to the Hard Problem of Consciousness." His more developed account is *The Conscious Mind.*
15. Chalmers, *The Conscious Mind,* 56–71, 93–118.
16. This synecdochal form of identification has been critiqued extensively in science studies literature. See, e.g., Doyle, *On Beyond Living.*
17. For recent arguments to this effect see Alaimo and Hekman, *Material Feminisms*; Kirby, *What If Culture Was Nature All Along?*; and Colebrook, "From Radical Representations to Corporeal Becomings."
18. Kirby, *What If Culture Was Nature All Along?,* 7. Hereafter cited parenthetically in the text.
19. Kirby is here discussing Wilson's *Gut Feminisms,* which focuses primarily on the hormones and microbiomes.
20. One might compare this to Hird's maneuver, following Margulis, to situate ourselves *as* the bacteria (see *Origins of Sociable Life*).
21. Keller's "bucket model" suggests that nature and culture are two contributing factors to an organism (*Mirage,* 9). We can reaffirm this in scalar terms: nature and culture usually point to different scales (atomic/cellular and social/ecological) that contribute to the creation of what is literally the same thing. The division between nature and culture is a mirage in the sense that it is viewing the same thing from two different scales.
22. This argument further clarifies how, as Wilson argues, "biology is not a synonym for determinism and sociality is not a synonym for transformative" (*Gut Feminisms,* 8).
23. Mol's *The Body Multiple* arises from the need to account for how medicine delimits the body in contradictory and troubling ways. She turns to a multiplicity to push against the tenuous unities that are so frequently invoked or assumed within medical practice (e.g., 84). Often the problem is remaining on one scale, as in when doctors "do not zoom out from the details under the skin to the patient as a whole, but move the camera

sideward and focus it on another object" (124). In this way, the turn to the fragmented and the multiple becomes a productive experiment in pushing against scalar preferences and occlusions within the medical profession, particularly as they work under various assumptions of coherent selves and others (see 129–34).

24. Hayles, *How We Became Posthuman*, 49.
25. Hayles, 25, 36. Although there is no room to elaborate on this argument, I'd suggest, to the contrary, that information allowed us to grant communication a body in a new way. Thus, Weaver's interpretation of Shannon's information theory provides a basis for articulating different scales at which information is instantiated and brought into effect (see Shannon and Weaver, *Mathematical Theory of Communication*, chapter 1).
26. Hayles, 287.
27. Keller and Grontowski, "Mind's Eye," 212.
28. See 186e; Keller and Grontowski, "Mind's Eye," 212.
29. Compare the discussion of the philosopher in 174b–175b, where Socrates discusses the philosopher's kind of knowledge in scalar terms.
30. Keller and Grontowski, "Mind's Eye," 215, emphasis added.
31. The interlocutors of *Theaetetus* state that we could, perhaps, posit that capacities of the mind "unite in one power, whether we should call it soul [*psukhēn*] or something else, by which we perceive through these as instruments the objects of perception" (184d). This turn to the *psukhē* is itself a turn away from *the perception alone* to the comparison and calculation of this perception: "The soul most certainly views their relations to one another" by "reflecting within itself" on the structure of these perceptions (186b).
32. *Manas* is more comparable to Heraclitus's notion of the Logos, as Aurobindo argues in *Kena and Other Upanishads*, 32. For an extensive examination of *cit*, see Gupta, *Cit*.
33. Remember that scalar shifts are not always equivalent to levels of abstraction. This final clause is a bigger claim than it initially appears and may not hold out upon inspection. After all, it suggests some strange things about our experience, especially that the scale at which one might reflect on experience is not the same as what can be best said to produce it.
34. Haraway, "Situated Knowledges," 585.
35. Emerson, "Nature," 10.
36. Mulvey, "Visual Pleasure and Narrative Cinema," 8.
37. Sontag, *On Photography*, 7, 42.
38. Welling, "Ecoporn," 53.

39. Merchant, *Reinventing Eden.*
40. See, e.g., Morton, *Ecology without Nature.*
41. Mulvey, "Visual Pleasure and Narrative Cinema," 8.
42. Haraway, "Situated Knowledges," 589. Hereafter cited parenthetically in the text.
43. Johnston, *The Cloud of Unknowing,* 124.
44. Nagel, *The View from Nowhere,* 63.
45. Nagel, 61.
46. See Haraway's "Cyborg Manifesto" (154), also discussed in Wolfe, *Critical Environments,* 44. Doyle (*On Beyond Living,* 106) has a critique of Haraway's location similar to what is articulated here.
47. The point here is not only conceptual but affects studies invoking the term "god trick," as I have discussed elsewhere (see "Scale Tricks and God Tricks"). There, I examine how scale confuses Messeri's invocation of the god trick in *Placing Outer Space,* 108.
48. See Shankara's *The Vivekachudamani,* e.g., "Transfer the 'I'-thought, which has sprung from the body, to the Self, which is consciousness and ever-blissful" (183).
49. Bourgine and Stewart, "Autopoiesis and Cognition," 327.
50. Maturana and Varela, *Autopoiesis and Cognition,* xiii. Hereafter cited parenthetically in the text.
51. "Our approach will be mechanistic: no forces or principles will be adduced which are not found in the physical universe" (Maturana and Varela, 75).
52. Wolfe, *What Is Posthumanism?,* 254. Wolfe is using Luhmann's "openness from closure" to avoid this problem, but we can ask again: does not the primary foreclosure of "openness from closure" arise from the assumption that this is a unit and that this unit is the "one who experiences"? Cf. Haraway's "sympoiesis" as another counterpoint to these problems with autopoiesis (*Staying with the Trouble,* chapter 3).
53. Foerster, *Beginning,* 35, 130.
54. Foerster, 48.
55. Foerster, 133.
56. A. Clark and Chalmers, "The Extended Mind."
57. Pilsch, *Transhumanism,* 39. Hereafter cited parenthetically in the text.
58. Merrell-Wolff, *Franklin Merrell-Wolff's Experience and Philosophy,* 316–17.
59. Aurobindo, *The Synthesis of Yoga,* 190.
60. This is the image used by Nicholas of Cusa in his *Vision of God*: "Thou appearest to me like the shadow following the movement of one that walketh; but 'tis because I am a living shadow and Thou the truth that I judge from the change of the shadow that the truth changeth also" (72).

61. Maharshi, *Be As You Are*, 47. Bracketed portion in original.
62. Maharshi, 49.
63. Maharshi, 49.
64. Rao and Sharma, *The Ribhu Gita*, chapter 22, verse 33 (378). Hereafter cited parenthetically using chapter and verse (e.g., 22.33).
65. Mitchell, *The Way of the Explorer*, 58.
66. 1 Corinthians 13:9–12, King James Version.

9. CUTTING AND CLAIMING EVERYTHING

1. Schweickart, "No Frames, No Boundaries."
2. Julian of Norwich, *Revelations of Divine Love*, 7.
3. Julian of Norwich, 8.
4. Sargeant, *Bhagavad Gita*, 11.13–15; translation modified.
5. These conversations and critiques of the space program in the 1960s have been extensively traced out in recent scholarship, e.g., in Poole, *Earthrise*, and Tribbe, *No Requiem for a Space Age*.
6. Arendt, *The Human Condition*, 1. Hereafter cited parenthetically in the text.
7. Cicero, *Somnium Scipionus*, 7.
8. Cicero, 10–11.
9. Cosgrove, *Apollo's Eye*, 5.
10. Cosgrove, 53.
11. F. White, *The Overview Effect*, 3.
12. Wells, *War of the Worlds*, 3.
13. Wells, 69.
14. F. White, *The Overview Effect*, 4.
15. Despite the recent work in philosophy of science, following Ian Hacking's *Representing and Intervening*, highlighting observation as an act of intervention, we need to hold them apart when considering the structure of the intervening.
16. F. White, *The Overview Effect*, 4.
17. Whitman, *Democratic Vistas*, 4. Hereafter cited parenthetically in the text.
18. Latour, *Reassembling the Social*: "It was at this point that nonhumans—microbes, scallops, rocks, and ships—presented themselves to social theory in a new way" (10). Hereafter cited parenthetically in the text.
19. Latour, *Inquiry*, 403–4.
20. Originally, the geographers Peter J. Taylor and Neil Smith had a simpler schematic that didn't include the body. As Herod notes, subsequent critiques led to an expansion of these divisions (*Scale*, chapter 1). For an overview see Howitt, "Scale"; Herod and Wright, *Geographies of Power*; and Herod, *Scale*.

21. Marston, Jones, and Woodward, "Human Geography without Scale"; Marston, "The Social Construction of Scale."
22. Latour, *Reassembling the Social*, 171. On the application of Lefebvre, see Marston, "The Social Construction of Scale," 227, and Herod, *Scale*, 26–32.
23. Marston, Jones, and Woodward, "Human Geography without Scale," 420 [hereafter cited parenthetically in the text]. Responses to this article, which are summarized in Herod, *Scale*, nonetheless keep some of the basic assumptions about the social construction of scale and the limited range of interest involved. See especially the discussion of the politics of scale in Herod, *Scale*, 14–15.
24. As an example of problematic interdisciplinarity, this discussion in geography has spread uncritically back into other disciplines in the humanities. M. T. Clarke and Wittenberg (*Scale in Literature and Culture*, 12–13); Brannon ("Standardized Spaces"), Tong ("Ecology without Scale"), Lukinbeal ("Scale"), and Carr and Lempert (introduction to their volume *Scale*) invoke and rely heavily on the geography conversation on scale for their arguments about other scalar representations. The result is that all these works set the conversation on scale within the limits provided by human geography.
25. Fraser, *Scales of Justice*, 2.
26. Fraser, 4.
27. The multitude is the key term for a similar critique, e.g. in Virno's *A Grammar of the Multitude*, which attempts to deny the synecdoche by reconceptualizing the aggregate as a multitude incapable of being reduced to a unity. Deleuze and Guattari are working in a similar vein, influenced by the Marxist anthropologist Pierre Clastres's *Society against the State*, which reassesses the assumption whereby leaders are given synecdochic authority over those they lead (see *A Thousand Plateaus*, especially chapter 12). DeLanda is deriving his assemblage theory directly from Deleuze and Guattari.
28. DeLanda, *Assemblage Theory*, 12. Hereafter cited parenthetically in the text.
29. The same point applies to Latour's notion of composing the collective as articulated in *Politics of Nature*. In a literal sense, the collective does not need to be composed; it is already together even as it continually rearranges itself (as the biosphere already does in these iterations of organisms that encompass the Earth). Even as we find ways to rearrange these relations, we need to acknowledge that these relations are, in some sense, internal to the Earth, and the problem is therefore as much a task of recognizing this structure, what we have left out, and what we already assume about it.

30. Harrington, *Reenchanted Science.*
31. Fraser, *Scales of Justice,* 130.
32. DeLanda, *Assemblage Theory,* 21.
33. Harrington, *Reenchanted Science,* 9. Hereafter cited parenthetically in the text.
34. The theoretical biologist Marcello Barbieri, in introducing his semantic biology, makes a similar point: "Why could ancient people think about atoms but not about cells?" (*The Organic Codes,* 17). I'd agree that "cell theory has undoubtedly been one of the greatest revolutions in the history of thought" (18), adding that it is so partially because it forces us to a more properly layered scalar conception of the world and of life. Sapp also makes a similar argument about the role of cell theory in evolution, writing that "the concept of beings radically changed in the middle of the nineteenth century. A plant or an animal was no longer conceived of as a singular entity, but comprised thousands of millions of mutually interdependent living units in a 'cell-commonwealth'" (*The New Foundations for Evolution,* 19–20).
35. A similar argument is made by Mazzarello in "A Unifying Concept."
36. Virchow quoted in Harrington, *Reenchanted Science,* 9.
37. See Harrington, 28–29. The part/whole question is entangled in the question of the machine, which seems to overtake theories of holism, at least in Harrington's account. But this concern about rendering life a machine is a separate issue that I think disappears in truly scalar mappings of relations such as those discussed in the next section.
38. On the reductionism dialectic see Bertalanffy, *Problems of Life,* 9–22, and *General Systems Theory,* 10–16. On his situating his theory in relation to summation and units see *Problems of Life,* 10.
39. Bertalanffy, *Problems of Life,* 11. It is essential that the "whole" invoked here is not the same as the Whole used throughout this book. In fact, chapter 7 would suggest that the word "whole" has created many problems in the history of biology and should be avoided.
40. Bertalanffy, 37.
41. Bertalanffy, 150.
42. Bertalanffy, 155.
43. Bertalanffy, *General Systems Theory,* 49.
44. Latour, *Politics of Nature,* 22: "Neither cybernetics nor hierarchies make it possible to understand the unbalanced, chaotic, Darwinian, sometimes local and sometimes global, sometimes rapid and sometimes slow agents."
45. Wiener, *Human Use,* 27.

46. Galison, "The Ontology of the Enemy," 263, 266.
47. Bateson, *Steps*, 485.
48. Bateson, 485.
49. Bertalanffy, *General Systems Theory*, 55.
50. Maturana and Varela, *Autopoiesis and Cognition*, 90–93. Hereafter cited parenthetically in the text.
51. On this disagreement see Hayles, *How We Became Posthuman*, 154.
52. See Luhmann, *Introduction to Systems Theory*, 68–70.
53. Luhmann, *Theories of Distinction*, 87.
54. Luhmann, 106.
55. See Maturana and Varela, *Autopoiesis and Cognition*, 96.
56. See Weaver's introduction to Shannon and Weaver, *The Mathematical Theory of Communication.*
57. West, *Scale*, 105. Hereafter cited parenthetically in the text.
58. West, 107: "Of equal importance was their appreciation that, to varying degrees, all theories and models are approximate."
59. Both Wolfram and Mandelbrot note how they are indebted to computer generation of images and iteration. See Wolfram, *New of Kind of Science*, 43–50; Mandelbrot, *The Fractalist*, 249.
60. Mandelbrot, *Fractal Geometry of Nature*, 3. Hereafter cited parenthetically in the text.
61. Kauffman, *At Home in the Universe*, 176. Hereafter cited parenthetically in the text.
62. This myth can be compared with Foerster's "little hooks" description in which you imagine a series of little hooks in a bin that you simply shake to form complex chains (*Beginning*, 34).
63. Once such emergence happens, a living system can then rewrite its own rules, a possibility once suggested by Pattee in "How Did a Molecule Become a Message?"
64. Wolfram's *New of Kind of Science* also deals with fractals (187–92) and networks (193–204).
65. This trope has been critiqued for the way it permits an obfuscation of the particularities of life and potentially an extension of notions of ideas of control (see Doyle, *On Beyond Living*, chapter 6; Hayles, *How We Became Posthuman*, chapter 9). These critiques of artificial life mirror the discussion of nanotechnology in the next section. The aim here is to resituate the practice more properly as a kind of abstract modeling of the possibilities of aggregation.
66. Wolfram, *New of Kind of Science*, chapters 2–6. Hereafter cited parenthetically in the text.

67. Mandelbrot suggests something similar in his discussion of the use value of the pictures of fractal sets (*Fractal Geometry of Nature*, 21–23).
68. Kauffman relies on Prigogine's notion of dissipative systems to accentuate this point (*At Home in the Universe*, 20–23). In *Order out of Chaos*, Prigogine and Stenger clarify this point in scalar terms, both in relation to the history of philosophy (8) and the modeling of dissipative systems both chemical (146) and biological (153).
69. Wolfram handles attractors in *New of Kind of Science*, 275–80.
70. For a summary of Margulis's role in Gaia theory see Bruce Clarke's "Neocybernetics of Gaia."
71. Margulis, *Symbiotic Planet*, 6. See also chapter 5 of Margulis and Sagan, *What Is Life?*
72. See Margulis, *Symbiotic Planet*, 117.
73. See Dorion Sagan and Lynn Margulis, "Gaia and Philosophy," in Margulis and Sagan, *Dazzle Gradually*, 181.
74. Margulis, *Symbiotic Planet*, 2.
75. Margulis and Sagan, *Microcosmos*, 125. Dorion notes that he and Margulis were given the contract for *Microcosmos* in the heyday of the premiering of *Cosmos* (Margulis and Sagan, *Dazzle Gradually*, 13).
76. Margulis often referenced the bacteriologist Sorin Sonea, whose *A New Bacteriology* (with Maurice Panisset) proposes that bacteria ought to, in some sense, be considered one enormous super-organism: "Bacteria have revealed that they form a highly efficient, complex, unified entity" (23).
77. Hird, *Origins of Sociable Life*, 26.
78. See Teilhard de Chardin, *The Phenomenon of Man*, 182–85; Vernadsky, "The Biosphere and the Noösphere"; Margulis and Sagan discuss the nöosphere in *What Is Life?* (170).
79. See Lovejoy, *Great Chain of Being*.
80. Many nanoscientists would like to claim Feynman but not Drexler, but both are essential for framing nanotechnology as a phenomenon. Milburn calls this the "Feynman origin myth" (*Nanovision*, 36).
81. Feynman, "Plenty of Room" 34.
82. Feynman, 34.
83. Drexler, *Engines of Creation*, 4. Hereafter cited parenthetically in the text.
84. Smalley, "Of Chemistry, Love, and Nanobots."
85. Published as Baum, "Nanotechnology." Hereafter cited parenthetically in the text.
86. Milburn's *Nanovision* considers the visionary practices of nanotechnology and science fiction. Doyle's *Wetwares* considers the anticipatory nature of nanotechnology (136–37).

87. Milburn, *Mondo Nano*, 61.
88. Kassem et al., "Stereodivergent Synthesis."
89. Haraway, *Staying with the Trouble;* Doyle, *Darwin's Pharmacy* (attention attractors at 37; rhetorical technologies at 22–23, 36).
90. Doyle, *Darwin's Pharmacy*, 7.
91. Plato describes a kind of flash of insight as the purpose of all his teaching in the seventh letter—see chapter 10.
92. Aurobindo, *The Synthesis of Yoga*, 43.
93. Pseudo-Dionysius, *Complete Works*, 154. Hereafter cited parenthetically in the text.
94. Huxley, *The Perennial Philosophy*, 11.
95. Arendt, *The Human Condition*, 20.
96. A potent example can be found in Hallward's critique of Deleuze's mysticism in *Out of This World*. Much of Deleuze's philosophy treads close to what we have discussed in this chapter; in this case the accusation of mysticism in Deleuze is welcome. However, Hallward's accusation that "the political aspect of Deleuze's philosophy amounts to little more than utopian distraction" and that "Deleuze's work is essentially indifferent to the politics of this world" (162) reiterates this trope that the contemplative is inherently antithetical to political transformation. In turn, Hallward's model of political change is noticeably nonscalar, such as in his suggestion that our "politics of the future" needs "more resilient forms of cohesion" (162), as if that unity has to be imposed within and out of political actions.
97. Bateson, *Mind and Nature*, 195.
98. Bateson, 199.
99. Whitman, *Song of Myself*, sections 31 and 32.

10. MAPPING THE VAST UNKNOWING

1. I am referring to the *Cosmos* series by their hosts: Sagan's *Cosmos: A Personal Voyage* and Tyson's *Cosmos: A Spacetime Odyssey.*
2. Sagan, *Cosmos*, episode 1.
3. Sagan, *Varieties of Scientific Experience*, ix.
4. Sagan refused to comment on postmodernism except in passing, but he also rebuffed attempts by his son Dorion to get him to read Derrida and others. See Dorion Sagan, "Truth of My Father," in Margulis and Sagan, *Dazzle Gradually*, 12.
5. This notion of the two cultures is one of the most persistent tropes of this conversation since C. P. Snow's 1959 lecture.
6. Latour, *Pandora's Hope*, 168–70.

7. From a historical perspective, it is not enough to reduce the discovery of the cell to the microscope (see Henry Harris's *The Birth of the Cell*, 1–3). Likewise, the microscope has always been subject to contention, which required us to understand as much about the eye as about the apparatus (see Schickore, *The Microscope and the Eye*). A conceptual switch had to occur, which we discussed in chapter 9. However, without the microscope such a conceptual switch would have never been possible.
8. Rheinberger, *Toward a History of Epistemic Things*, 28.
9. Daston and Galison document this process for microscopy in *Objectivity*.
10. See Messeri's discussion of the relationship between planetary astronomy and various models (*Placing Outer Space*); Myers's discussions of protein modeling in *Rendering Life Molecular*; and Hanson's examination of imagery in nanotechnology in *Haptic Visions*. Perhaps the most widely discussed form of such diagrams are those related to global warming (see, e.g., the review essays by S. J. O'Neill and Smith, "Climate Change and Visual Imagery," and Walsh, "The Visual Rhetoric of Climate Change").
11. See Myers, *Rendering Life Molecular*, 56.
12. On metascientific models see Callon, "Four Models." Note that I have deployed this somewhat awkward use of "specification" at crucial locations in Part I: 5.1, 5.5, 5.7, 5.11, 6.25–26, and 6.43.
13. If we combine this notion with the third thought experiment in scale (chapter 6) we get a playfully videogame-like idea in which you could imagine selecting any object within the interface to see all its stats and relations.
14. Here, I am thinking somewhat about Latour's notion of the black box as the hiding of the conditions for truth creation within facts and technologies, particularly as is exemplified by his opening examples in *Science in Action*. However, this term "black box" has significant rhetorical problems, in that it implies that these forms of specification are truly hidden and inaccessible.
15. This relationship between the epistemic thing and the experimental system is the primary frame for Rheinberger's *Toward a History of Epistemic Things*.
16. Barad's notion of the apparatus in *Meeting the Universe Halfway* (discussed in chapter 7) would be applicable here, as would Andrew Pickering's assessment of the machinic performances of scientific practice in *The Mangle of Practice*.
17. Deleuze and Guattari, *What Is Philosophy?*, 118, 117.
18. Deleuze and Guattari, 49.

19. Heidegger, "Science and Reflection," in *The Question concerning Technology,* 168–69.
20. Heidegger, 173.
21. This claim has been made by Latour in earlier work but is discussed in frank terms in "Why Has Critique Run out of Steam?" and "Coming Out as a Philosopher."
22. The opening reflections in Bloor's *Knowledge and Social Imagery* (1–5) and in Latour and Woolgar's *Laboratory Life* (17–27) provide fascinating windows into the kinds of justifications and resistances they faced at the outset.
23. In the opening reflections of Harding, *The Science Question in Feminism*: "The radical feminist position holds . . . that the social structure of science, many of its applications and technologies, its modes of defining research problems and designing experiments, its ways of constructing and conferring meanings are not only sexist but also racist, classist, and culturally coercive" (9). The conversation about research into women's bodies and reproduction is particularly potent on this front. For example, Roberts, *Killing the Black Body*.
24. Latour and Woolgar speak of "an effort . . . to penetrate the mystique of science" (*Laboratory Life,* 18). See Hacking on Karl Mannheim's distinction between unmasking and refuting (*Social Construction,* 53–58).
25. Bloor, *Knowledge and Social Imagery*: "Of course knowledge must be distinguished from mere belief. This can be done by reserving the word 'knowledge' for what is collectively endorsed, leaving the individual and idiosyncratic to count as mere belief" (3). See also Bloor, "Anti-Latour," 109. There is, of course, a great deal of variety in the strength of the claims prioritizing certain aspects such as the social.
26. Bloor, *Knowledge and Social Imagery,* 4–5.
27. Hacking's analysis of the notion of "social construction" in *The Social Construction of What?* might be usefully situated in this way. Clearly, much of the problem of, for example, Pickering's analysis of quarks or Latour and Woolgar's analysis of thyrotopin-releasing hormone is the attitude of unmasking (see Hacking, *Social Construction,* 94), but we can also think of the problem as about confusion of the scale of specification—in what is taken as the primary determinant.
28. The attempt to realize aspects of agency beyond humans is one that I am obviously sympathetic to, as is clear in Part II and my turn to mysticism. But Latour has done so by extending the language of human agency and relations to the world more generally, rather than starting with a

consideration of the notion and form of agency, which would have led to a very different project.

29. Latour, "An Attempt at a 'Compositionist Manifesto,'" 478.
30. Latour, *Pandora's Hope*, 168–70.
31. Latour, "One More Turn," 283.
32. Latour, "Reply," 123.
33. Bloor, "Anti-Latour": "Latour, however, treats [Shapin and Schaffer's *Leviathan and the Air-Pump*] as an account of the emergence of the subject-object schema itself, rather than an account of one of the historical and cultural forms that our understanding of it, as a biological given, has assumed" (107).
34. See Latour, "Coming Out as a Philosopher," 601–2.
35. Latour, "Why Has Critique Run out of Steam?," 232.
36. A more extended attempt to do so can be found in Felski's *The Limits of Critique*.
37. Latour, "An Attempt at a 'Compositionist Manifesto,'" 475.
38. Latour, 475.
39. Latour, 476.
40. There is some overlap here with Latour's notion of perplexity articulated in *Politics of Nature* (104), but the process he describes there remains firmly within his political language. But consider: what is the primary affect one takes when entering a political frame for action? Does it allow one to pause and dwell in this way?
41. Latour, "Why Has Critique Run out of Steam?," 232.
42. Sagan, like many others, attempts a more balanced approach, noting that science "invites us to let the facts in, even when they don't conform to our preconceptions" and that it "urges us to a delicate balance between no-holds-barred openness to new ideas, however heretical, and the most rigorous skeptical scrutiny of everything" (*Demon-Haunted World*, 30).
43. Sagan, 6–7.
44. See Bolinsky, "Inner Life of the Cell."
45. Plenty of critical theorists have approached science in similar ways but with more or less bewilderment and sense of wondering. Jakobson's and Levi-Strauss's enthusiasm for information theory, the long-standing uses of cybernetics, systems theory, and chaos theory in Deleuze, Lyotard, Donna Haraway, Isabelle Stengers, Cary Wolfe (the list could continue); the invocations of biological inscriptions by Derrida (see Vitale, *Biodeconstruction*); the uses of set theory by Badiou, relativity in Virilio, quantum physics in Barad and Pickering, symbiogenesis in Bruce Clarke and Myra

Hird—all of these are born out of an attempt to take seriously the productions of science, to see how these ways of configuring reality function, translate, imply, bewilder, disrupt, confuse.

46. Polanyi, *Tacit Dimension,* 34.
47. Polanyi, 34.
48. Walls, *The Passage to Cosmos.* Hereafter cited parenthetically in the text.
49. Humboldt, *Cosmos,* ix. Hereafter cited parenthetically in the text.
50. Three potential examples come to mind: Derrida's philosophy read not simply in the critical sense we just discussed but as a kind of sitting with the effects of language (see especially "Signature Event Context" in *Limited Inc.*); Heidegger's notion of dwelling; and, perhaps most directly, Kierkegaard's willingness to sit in extended contemplation even with angst. But each of these arrives closest to our questions here when they are also the most aligned with mysticism—somewhat obviously in Kierkegaard, but also in Derrida's similarity with negative theology (see his discussion in "How to Avoid Speaking" in *Psyche: Volume II*) and in Heidegger's relationship to Meister Eckhart.
51. The same could be said of poetics given the recurring mysticism of major poetic movements, from Dante to Blake to Whitman in English, and more explicitly in the epic traditions of India and in the mystical poetics of figures like Kabir, Rumi, and Tagore.
52. Theodore Schick Jr. and Lewis Vaugh, *How to Think about Weird Things,* quoted in Sagan, *Demon-Haunted World,* 234. Sagan comments on this later in relation to "postmodernist texts" (284).
53. Sagan, *Demon-Haunted World,* 238.
54. Gross, *The Rhetoric of Science,* 33.
55. Gross and Keith, introduction to *Rhetorical Hermeneutics,* 1. Rhetoricians have continued to draft Latour into their conversations, which he has since expressed some resistance to perhaps on the basis of these earlier conflations. See Walsh et al., "Forum."
56. For a summary of these reactions see Harris, "Alan Gross," and W. J. White, "Disciplinarity."
57. Republished in Gross and Keith, *Rhetorical Hermeneutics.* This reprint is cited hereafter parenthetically in the text.
58. Responses to Gaonkar were published as Gross and Keith, *Rhetorical Hermeneutics,* 1–24. Gross's revision was retitled *Starring the Text.*
59. Harris ("Alan Gross") points to Fahnestock's *Rhetorical Figures of Science,* Condit's *The Meanings of the Gene,* and Ceccareli's *Shaping Science with Rhetoric* as examples. Many other studies by those trained in rhetoric have

been released since, which take a variety of approaches to science. See, e.g., Teston, *Bodies in Flux*; Johnson, *American Lobotomy;* Ceccarelli, *On the Frontier of Science;* Doyle, *Darwin's Pharmacy*; Hanson, *Haptic Visions;* Graham, *The Politics of Pain Medicine*—just to name a few. Such studies are united by the disciplinary distinction of rhetoric in the U.S academy, by a variety of methods geared toward drawing out and analyzing how communication works in science (see, e.g., Rai and Druschke, *Field Rhetoric*) and by some grounding in the longer traditions of rhetoric.

60. Versluis, *Platonic Mysticism*, traces the origin of the term *mysticism* to Platonic traditions; he is less clear on whether the term ought to be applied directly to Plato, but others—in a more critical vein—have done so with ease (see, e.g., Russell, "Mysticism and Logic"), and we can accept, at the least, that Plato was read this way through the Neoplatonists, Pseudo-Dionysius, and into the Christian traditions as it is expounded later by Evelyn Underhill and others. On the origin of the term *rhetoric*, Schiappa has, with some controversy, claimed that Plato coined the term in the *Gorgias*, deriving it from the previous term *rhetor* as the speaker in the Assembly (see Schiappa, "Did Plato Coin *Rhētorikē*?"). Regardless of the validity of this claim, Plato's notion of rhetoric, particularly in the *Gorgias* and the *Phaedrus*, has influenced all discussions of the term after him.
61. For an extensive overview of this history of setting up Plato as the enemy of rhetoric see Jasso, "Sympathy for the Devil."
62. See discussion of "Knowledge through Identity" in chapter 7.
63. Contrast these with Socrates's repeated statements about himself as "knowing my own ignorance" (*Phaedrus* 235c) and "For the things I say I certainly don't say with any knowledge at all" (506a). It is interesting that at these moments we often want to say that Socrates is being ironic, because we assume that Socrates surely does know the answer and is trying to merely catch his opponents. But this is precisely what is performed between Callicles and Socrates—even down to the accusation of irony at 489e.
64. See Lidell-Scott-Jones, "doxa."
65. See, e.g., 471e, 473e, 489b, 494d.
66. Sagan, *Billions and Billions*, 10.
67. Sagan, 10.

11. THE COSMOS SEEING ITSELF

1. This quote is from Brand's recounting of the experience as reposted to his personal website as "Whole Earth Origin."

2. Poole, *Earthrise,* discusses this passage from Brand (150) and the impact of Brand's campaign (73).
3. See, e.g., Poole, *Earthrise,* chapter 8; Henry and Taylor, "Re-thinking Apollo," 193; Cosgrove, *Apollo's Eye,* 257.
4. On the commodification of the Earthrise and Blue Marble photos see Poole, *Earthrise,* 168; Tribbe, *No Requiem for a Space Age,* chapter 5.
5. Helmreich addresses these readings in "From Spaceship Earth to Google Ocean," 1214–15. His reading is a notable exception to the generally negative readings. Most of these negative readings bring the picture back to its political implications then or since. One of the most thorough is McGuirk, "A. R. Ammons and the Whole Earth." See also Tong, "Ecology without Scale," 197 (Figure 1 caption); Russill, "Earth-Observing Media"; Lazier, "Earthrise"; Gurevitch, "The Digital Globe"; Geppert, "Rethinking the Space Age"; DeLoughrey, "Satellite Planetarity," 261; Kurgan, *Close Up at a Distance,* 41. Two readings that note and then move beyond the colonial reading include Kelsey, "Reverse Shot," and Messeri, *Placing Outer Space,* 194.
6. Garb, "Use and Misuse," 18.
7. Garb, 18.
8. Most theories of scale used for cultural criticism are derived from geographers (see chapter 9, endnote 24). Another approach has been through Wallerstein's world-systems theory; this was central to the special issue of *Modernism/Modernity,* e.g., as engaged by Vadde's "Scalability" (note that the geography conversations on scale have been infused with Wallerstein since the 1980s [see Herod, *Scale,* 8]). Some work has usefully cited ecologists (e.g., T. Clark, "Scale"; Woods, "Scale Critique," 136), but as M. T. Clarke and Wittenberg note, much of this work remains focused on the large scale (introduction, 24). Thus, the theories of scale used in these works remain limited, with two notable exceptions: Horton's "Composing a Cosmic View" situates scale in relation to Paracelsus and Whitman, offering a broader view of scalar productions; and Woods's "Epistemic Things" considers Rheinberger's epistemic things in relation to objects from other scale domains.
9. I use "critical" here more in the negative, pejorative sense rather than the more formal sense developed in the last chapter.
10. For example, in "Epistemic Things," Woods invokes Haraway and Arendt. Horton, "Composing a Cosmic View," invokes Cosgrove.
11. Garb, "Use and Misuse," 18. See Lakoff and Johnson, *Metaphors We Live By,* 5.
12. Garb, "Use and Misuse," 18.

13. Garb, 18.
14. Garb, 18. Helmreich ("From Spaceship Earth to Google Ocean," 1213) appropriately talks about the image of Earth as an icon or totem, using Peirce's notion of the icon.
15. Woods, "Scale Critique."
16. Tsing, "On Nonscalability" and *The Mushroom*, chapter 3.
17. Tsing, "On Nonscalability," 508.
18. Tsing's case study here is the matsutake mushroom, which she develops at length in the beautiful ecological-economic contemplation *The Mushroom at the End of the World.*.
19. Tsing, "On Nonscalability," 505.
20. Chakrabarty, "The Climate of History," 206–7.
21. Many scholars have critiqued the focus on the human in this term, since it seems to make the epoch about the human; see, e.g., Haraway, *Staying with the Trouble*, 2.
22. Woods, "Scale Critique," 133.
23. T. Clarke, "Scale," 151.
24. For this point applied to environmental communication see DiCaglio, Barlow, and Johnson, "Rhetorical Recommendations."
25. Kurgan, *Close Up at a Distance.* See also DeLoughrey, "Satellite Planetarity." Kurgan begins the book citing Cosgrove's critiques of Earth imagery in *Apollo's Eye* (discussed here in chapter 9). This kind of critique is also in line with Haraway's observation of the "Land ho!" phenomenon, in which space researchers put pictures of planets on their walls in a manner comparable to colonial lands as objects to be dominated (*Simians, Cyborgs, and Women*, 221).
26. Kurgan, *Close Up at a Distance*, 26.
27. Kurgan, 30.
28. Such metaphors are often the basis for critiques of scale, e.g., in Tong, "Ecology without Scale," 199; Herod and Wright, *Geographies of Power*, 7.
29. An earlier version was produced by Eva Szasz in 1968 (*Cosmic Zoom*) using a combination of images and animation, but *Powers of Ten* is noteworthy for using photographs. Sagan's and Tyson's versions of *Cosmos* both make use of this same zoom.
30. For historical analysis of the production of *Power of Ten* see Woods "Scale Critique" and "Epistemic Things"; and Horton, "Composing a Cosmic View."
31. See, e.g., Woods, "Scale Critique" and "Epistemic Things"; Di Palma, "Zoom"; Latour, "Anti-zoom"; Tung, "Baddest Modernism"; Brannon,

"Standardized Spaces"; Tong, "Ecology without Scale"; Kurgan, *Close Up at a Distance,* 18–19. Horton, "Composing a Cosmic View" preferentially examines earlier, non-film depictions. I have overviewed and addressed these critiques in "Scale Tricks and God Tricks."

32. Woods, "Scale Critique," 63.
33. The same point is made in Latour, "Anti-zoom"; Clark, "Scale"; Horton, "Composing a Cosmic View," 53; and Woods, "Scale Critique," 134.
34. Di Palma, "Zoom," 260.
35. Latour, "Anti-zoom," 98.
36. See Woods, "Scale Critique," 134, and "Epistemic Things," 62.
37. In contrast, other depictions have attempted to confuse this view of scale, to different ends. Latour's "Anti-zoom" was written to accompany a 2014 exhibit by Olafur Eliasson at the Fondation Louis Vuitton in Paris. Latour here wants to speak of connectivity but not scale, since he associates scale with hierarchy and power. Thus he describes Eliasson's aim as a disorientation ("Anti-zoom," 101). In response, we can note that our account of scale shows that disorientation is not antithetical to scale but essential for its reorientation. As another contrast, the scalar exhibit by Stephen S. Young and Paul Kelly, *Macro or Micro?,* deliberately mixes up scales, placing together large-scale and small-scale imagery that looks similar in order to blur the differences in scale (see Young and Kelly, "Macro or Micro?," 253). Such blurring can be useful for identifying isomorphisms across scales and for showing how strange scalar views can be, but it will not teach us about the apparatus of scale itself that helps us account for these different perspectives.
38. Woods, "Epistemic Things," 78.
39. Woods, 78.
40. Brannon, "Standardized Spaces," 280.
41. Woods, "Scale Critique," 135.
42. Primack and Abrams, *The View from the Center,* 173–74; see also 4.18 here.
43. See also the discussion of *sūnyata* in chapter 7.
44. A great deal has been made of the technological requirements for these photos to be produced, particularly the reconstructed nature of them in the processing from film, the adding of color, and even the reorientation of them to fit our normal way of looking at the Earth. See, e.g., Dodge and Perkins, "The 'View from Nowhere'?"; Deese, "The Artifact of Nature"; DeLoughrey, "Satellite Planetarity"; Olwig, "The Earth Is Not a Globe." Many of these critiques extend a larger conversation about nature photography, e.g., Deluca and Demo, "Imaging Nature."

45. See Tribbe, *No Requiem for a Space Age,* 116, 133.
46. Whitman, *Song of Myself,* section 52.

12. TRANSFORMATIONS BY INVOLUTION

1. J. B. Taylor, "My Stroke of Insight."
2. The psychiatrist Walter Pahnke used the term *ego death* as an explicit bridge between religious and psychedelic experience (Pahnke, "The Psychedelic Mystical Experience"). The term has since remained important for research on the effects of psychedelics. See Doyle's discussion of ego death in *Darwin's Pharmacy,* 112, and "Healing with Plant Intelligence," 30–31.
3. Aurobindo, *The Synthesis of Yoga,* 189.
4. Aurobindo, 189.
5. The term *involutionary* can be found in both Aurobindo (e.g., *The Life Divine,* 122) and Deleuze and Guattari (e.g., *A Thousand Plateaus,* 238). Doyle and Connor have used the term *involutionary rhetoric* in this way in "Chanting the Supreme Word."
6. Wolfe, *What Is Posthumanism?,* 122.
7. Littlefield and Johnson, *Neuroscientific Turn,* 5–8.
8. Illes, preface to Littlefield and Johnson, *The Neuroscientific Turn,* xii.
9. Libet, "Unconscious Cerebral Initiative." For an overview of more recent studies testing Libet's original findings, see Soon et al., "Unconscious Determinants."
10. Massumi, *Parables for the Virtual,* 195.
11. Bennett and Hacker call this ascription of attributes to the brain the "merelogical fallacy" of neuroscience, since it attempts to ascribe attributes derived from the whole human body to a part of that body (*Philosophical Foundations of Neuroscience,* 68–74). But their articulation is largely situated in this-scale terms: selecting one object on this scale (the brain) to stand in for another (the human organism). The basic point holds, perhaps, but thinking of the shift as body-brain to neuron-experience would alter their account, since this shift becomes two ways of describing the same thing rather than synecdochically selecting one part (brain) to fill in for something (body). Dennett's response to Bennett and Hacker (reprinted in *Neuroscience and Philosophy*) provides a useful overview of his attempt to find other *levels* (what he calls the "sub-personal" ["Philosophy as Naive Anthropology," 77–79]) for describing attributes of mind. From my limited view, I only want to offer the question: how are these levels made to relate to each other if we understand them as two ways of

describing the same thing, at one point as an organism with a cognitive operation and at another point as neurons processing this cognition?

12. See Rotman, *Becoming Besides Ourselves,* and A. Clark and Chalmers, "The Extended Mind."
13. Soon et al., "Unconscious Determinants," 543.
14. The following discussion was greatly assisted by Gary Weber, who has been involved in many of the studies discussed. For a concise synthesis of the neuroscience research and its relation to thoughts, see Weber's "What 'No Thoughts' Means—3 Different Kinds of Thoughts." Weber's involvement in recent neuroscience of meditation is discussed in Robert Wright's *Why Buddhism Is True,* 167–71. This picture of the DMN is also described by Goleman and Davidson's in *Altered Traits,* 149–53.
15. Raichle et al., "A Default Mode of Brain Function."
16. Andrews-Hanna et al., "Functional-Anatomic Fractionation," 550.
17. Buckner, Andrews-Hanna, and Schacter, "The Brain's Default Network," 1. The DMN was highlighted as one of the major effects to be seen in neuroimaging studies in a meta-analysis by K. C. R. Fox et al. ("Functional Neuroanatomy of Meditation").
18. For example, the review article by Li, Mai, and Liu, "The Default Mode Network," surveys how the DMN has been connected to social understanding of others. However, early studies also found a connection between DMN and depression, e.g., Sheline et al., "The Default Mode Network."
19. Farb et al., "Attending to the Present."
20. V. A. Taylor et al., "Impact of Meditation Training," 11.
21. Brewer et al., "Meditation Experience."
22. For an excellent discussion of the scope of these scientific claims see Weber and Doyle, "The Neuroscience of Non-dual Awakening," in *Into the Stillness,* 189.
23. M. D. Fox et al., "The Human Brain," 9675; Vanhaudenhuyse et al., "Two Distinct Neuronal Networks"; Hamilton et al., "Default-Mode and Task-Positive Network Activity"; Uddin et al., "Functional Connectivity."
24. Hathaway, "Tuning Out."
25. See Garrison et al., "Real-Time fMRI."
26. Recently, more attention has been turned to the challenges, both methodological and conceptual, of these neuroscientific studies, e.g., in Tang, Hölzel, and Posner, "The Neuroscience of Mindfulness Meditation"; Van Dam et al., "Mind the Hype"; and Vieten et al., "Future Directions in Meditation Research." These are all constructively critical, indicating that

researchers have moved far beyond the domain of enthusiastic individuals pursuing research on a marginal topic.

27. Nearly everyone writing on mysticism in the last few decades has had to address this debate. For an overview see Byrne, "Mysticism, Identity, and Realism"; R. Leonard, *Transcendental Philosophy of Franklin Merrell-Wolff,* 78–91; Versluis, *Platonic Mysticism.*
28. Stace, *Mysticism and Philosophy;* Forman, *The Problem of Pure Consciousness.*
29. Katz, "Language, Epistemology, and Mysticism." See also Byrne, "Mysticism, Identity, and Realism," 237; King, *Orientalism and Religion,* 168.
30. Jantzen, *Power, Gender, and Christian Mysticism,* 337.
31. King, *Orientalism and Religion,* 73.
32. King, 75.
33. Augustine, *Confessions,* 68. Hereafter cited parenthetically in the text.
34. Cf. Julian of Norwich, *Revelations of Divine Love,* 19: "when the soul wills what God wills, it is then in a state like God."
35. In Bateson, *Steps.* Hereafter cited parenthetically in the text.
36. For a diagnosis of this use of identity as a means of biopower see Nealon, *Post-postmodernism.*
37. Foucault, *Hermeneutics of the Subject,* 30.
38. Foucault, *Technologies of the Self,* 18.
39. See Maharshi, *Who Am I?* My understanding of Maharshi has been greatly assisted by David Godman's assembled dialogues in *Be as You Are* and by the recordings of Robert Adams, one of Maharshi's proponents, available at robert-adams.info. Note that this emphasis and practice are by no means exclusive to Maharshi, but is a recurring practice and recommendation in the clearest contemplative texts.
40. Foucault, *Hermeneutics of the Subject,* 57.
41. This chant was passed to me orally by Richard Doyle, who learned it from Gary Weber, who has published his own translation and analysis in *Happiness beyond Thought,* 110–19.
42. Merrell-Wolff, *Transformations in Consciousness,* 130. Hereafter cited parenthetically in the text.
43. Merrell-Wolff, *Franklin Merrell-Wolff's Experience and Philosophy,* 6.
44. This is used as a repeated refrain in the Ribhu Gita, 14.5–25 (Rao and Sharma, *The Ribhu Gita*).
45. Johnston, *Cloud of Unknowing,* 74.
46. Johnston, 75. Cf. Merrell-Wolff on the "fire" in *Franklin Merrell-Wolff's Experience and Philosophy,* 36, and Julian of Norwich's experience with bliss and love (*Revelations of Divine Love,* 14–15).

47. Huxley, *The Perennial Philosophy,* 72; Mitchell, *The Way of the Explorer,* 66.
48. Merrell-Wolff, *Franklin Merrell-Wolff's Experience and Philosophy,* 117.
49. Johnston, *Cloud of Unknowing,* 117.
50. See Underhill, *Mysticism,* part 2, chapter 2.
51. Kierkegaard, *Fear and Trembling,* 83.

Bibliography

Ahl, Valerie, and T. F. H. Allen. *Hierarchy Theory: A Vision, Vocabulary, and Epistemology.* New York: Columbia University Press, 1996.

Alaimo, Stacy, and Susan Hekman, eds. *Material Feminisms.* Bloomington: Indiana University Press, 2008.

Andrews-Hanna, Jessica R., Jay S. Reidler, Jorge Sepulcre, Renee Poulin, and Randy L. Buckner. "Functional-Anatomic Fractionation of the Brain's Default Network." *Neuron* 65, no. 4 (February 25, 2010): 550–62. https://doi.org/10.1016/j.neuron.2010.02.005.

Arendt, Hannah. *The Human Condition.* 2nd ed. Chicago: University of Chicago Press, 1998.

Augustine. *The Confessions of Saint Augustine.* Translated by John K. Ryan. Garden City, N.Y.: Image, 1960.

Aurobindo, Sri. *Isha Upanishad.* 5th ed. Pondichery: Sri Aurobindo Ashram Press, 2014.

———. *The Life Divine.* Pondichery: Sri Aurobindo Ashram Press, 2010.

———. *The Synthesis of Yoga.* Pondichery: Sri Aurobindo Ashram Press, 1999.

Badiou, Alain. *Being and Event.* Translated by Oliver Feltham. London: Continuum-Bloomsbury, 2007.

Barad, Karen. *Meeting the Universe Halfway: Quantum Physics and the Entanglement of Matter and Meaning.* Durham: Duke University Press, 2007.

Barbieri, Marcello. *The Organic Codes: An Introduction to Semantic Biology.* Cambridge: Cambridge University Press, 2003.

Bateson, Gregory. *Mind and Nature: A Necessary Unity.* New York: Dutton, 1979.

———. *Steps to an Ecology of Mind: Collected Essays in Anthropology, Psychiatry, Evolution, and Epistemology.* Chicago: University of Chicago Press, 2000.

Baum, Rudy. "Nanotechnology." *Chemical and Engineering News* 81, no. 48 (December 1, 2003): 37–42.

Bennett, M. R., and P. M. S. Hacker. *Philosophical Foundations of Neuroscience.* Malden, Mass.: Wiley-Blackwell, 2003.

Bertalanffy, Ludwig von. *General System Theory: Foundations, Development, Applications.* Rev. ed. New York: George Braziller, 1969.

———. *Problems of Life: An Evaluation of Modern Biological Thought.* New York: Wiley, 1952.

Bloor, David. "Anti-Latour." *Studies in History and Philosophy of Science Part A* 30, no. 1 (March 1, 1999): 81–112. https://doi.org/10.1016/S0039-3681(98)000 38-7.

———. *Knowledge and Social Imagery.* London: Routledge, 1976.

Bohm, David. "Meaning and Information." In *The Search for Meaning: The New Spirit in Science and Philosophy,* edited by Paavo Pylkkanen, 43–85. Wellingborough, UK: Crucible, 1989.

———. *Wholeness and the Implicate Order.* London: Routledge, 2002.

Bohm, David, and Basil J. Hiley. *The Undivided Universe: An Ontological Interpretation of Quantum Theory.* London: Routledge, 1995.

Bolinsky, David. "Inner Life of the Cell." YouTube video, XVIVO Scientific American, 3:12, July 11, 2011. https://www.youtube.com/watch?v=wJyUtbnoO5Y.

Bonner, John Tyler. *Why Size Matters: From Bacteria to Blue Whales.* Princeton: Princeton University Press, 2011.

Bourgine, Paul, and John Stewart. "Autopoiesis and Cognition." *Artificial Life* 10, no. 3 (June 1, 2004): 327–45. https://doiorg/10.1162/1064546041255557.

Brand, Stewart. "Whole Earth Origin." Accessed April 25, 2021. http://sb.long now.org/SB_homepage/WholeEarth_buton.html.

Brannon, Monica M. "Standardized Spaces: Satellite Imagery in the Age of Big Data." *Configurations* 21, no. 3 (2013): 271–99. https://doi.org/10.1353con .2013.0021.

Brewer, Judson A., Patrick D. Worhunsky, Jeremy R. Gray, Yi-Yuan Tang, Jochen Weber, and Hedy Kober. "Meditation Experience Is Associated with Differences in Default Mode Network Activity and Connectivity." *Proceedings of the National Academy of Sciences* 108, no. 50 (December 13, 2011): 20254–59. https://doi.org/10.1073/pnas.1112029108.

Bricmont, Jean. *Quantum Sense and Nonsense.* New York: Springer, 2017.

Bucke, Richard Maurice. *Cosmic Consciousness.* Philadelphia: Innes and Sons, 1901.

Buckner, Randy L., Jessica R. Andrews-Hanna, and Daniel L. Schacter. "The Brain's Default Network." *Annals of the New York Academy of Sciences* 1124, no. 1 (2008): 1–38. https://doi.org/10.1196/annals.1440.011.

Byrne, Peter. "Mysticism, Identity, and Realism: A Debate Reviewed." *International Journal for Philosophy of Religion* 16, no. 3 (January 1, 1984): 237–43. https://doi.org/10.1007/BF00162669.

Cadava, Eduardo, Peter Connor, and Jean-Luc Nancy, eds. *Who Comes after the Subject?* New York: Routledge, 1991.

Callon, Michel. "Four Models for the Dynamics of Science." In *Handbook of Science and Technology Studies,* edited by Ulrike Felt, Rayvon Fouché, Clark A. Miller, and Laurel Smith-Doerr, 29–63. Rev. ed. Thousand Oaks, Calif.: SAGE Publications, 1995. https://doi.org/10.4135/9781412990127.

Carr, E. Summerson, and Michael Lempert, eds. *Scale: Discourse and Dimensions of Social Life.* Berkeley: University of California Press, 2016.

Ceccareli, Leah. *Shaping Science with Rhetoric.* Chicago: University of Chicago Press, 2001.

Chakrabarty, Dipesh. "The Climate of History: Four Theses." *Critical Inquiry* 35, no. 2 (2009): 197–222. https://doi.org/10.1086/596640.

Chalmers, David J. *The Conscious Mind: In Search of a Fundamental Theory.* Rev. ed. New York: Oxford University Press, 1997.

———. "Facing up to the Hard Problem of Consciousness." *Journal of Consciousness Studies* 2, no. 3 (1995): 200–219.

Cicero, Marcus Tullius. *Somnium Scipionis, the Dream of Scipio Africanus Minor, Being the Epilogue of Cicero's Treatise on Polity.* Translated by W. D. Pearman. Cambridge: Deighton, 1883. http://archive.org/details/somniumscipionis ooiciceuoft.

Clark, Andy, and David J. Chalmers. "The Extended Mind." *Analysis* 58, no. 1 (1998): 7–19.

Clark, Timothy. "Scale: Derangements of Scale." In *Telemorphosis: Theory in the Era of Climate Change,* vol. 1, edited by Tom Cohen, 148–66. Ann Arbor: Open Humanities Press, 2012.

Clarke, Bruce. "Neocybernetics of Gaia: Emergence of Second-Order Gaia Theory." In *Gaia in Turmoil: Climate Change, Biodepletion, and Earth Ethics in an Age of Crisis,* edited by Eileen Crist and H. Bruce Rinker, 293–314. Cambridge: MIT Press, 2010.

Clarke, Michael Tavel, and David Wittenberg, eds. *Scale in Literature and Culture.* New York: Palgrave Macmillan, 2017.

Clastres, Pierre. *Society against the State: Essays in Political Anthropology.* Translated by Robert Hurley and Abe Stein. New York: Zone Books, 1989.

Colebrook, Claire. "From Radical Representations to Corporeal Becomings: The Feminist Philosophy of Lloyd, Grosz, and Gatens." *Hypatia* 15, no. 2 (May 1, 2000): 76–93. https://doi.org/10.1111/j.1527-2001.2000.tb00315.x.

Condit, Celeste Michelle. *The Meanings of the Gene: Public Debates about Human Heredity.* Madison: University of Wisconsin Press, 1999.

Cosgrove, Denis. *Apollo's Eye: A Cartographic Genealogy of the Earth in the Western Imagination*. Baltimore: Johns Hopkins University Press, 2003.

Daston, Lorraine J., and Peter Galison. *Objectivity*. New York: Zone Books, 2010.

Davis, Thomas, and Nathan Hensley. "Scale and Form; or, What Was Global Modernism?" *Modernism/Modernity* 2, no. 4 (January 4, 2018). https://doi.org/10.26597/mod.0033.

Deese, R. S. "The Artifact of Nature: 'Spaceship Earth' and the Dawn of Global Environmentalism." *Endeavour* 33, no. 2 (June 2009): 70–75. https://doi.org/10.1016/j.endeavour.2009.05.002.

DeLanda, Manuel. *Assemblage Theory*. Edinburgh: Edinburgh University Press, 2016.

Deleuze, Gilles, and Félix Guattari. *A Thousand Plateaus: Capitalism and Schizophrenia*. Translated by Brian Massumi. Minneapolis: University of Minnesota Press, 1997.

———. *What Is Philosophy?* Translated by Hugh Tomlinson and Graham Burchell. Rev. ed. New York: Columbia University Press, 1996.

DeLoughrey, Elizabeth. "Satellite Planetarity and the Ends of the Earth." *Public Culture* 26, no. 2 (March 20, 2014): 257–80. https://doi.org/10.1215/08992363-2392057.

Deluca, Kevin Michael, and Anne Teresa Demo. "Imaging Nature: Watkins, Yosemite, and the Birth of Environmentalism." *Critical Studies in Media Communication* 17, no. 3 (2000): 241–60. https://doi.org/10.1080/15295030009388395.

Dennett, Daniel C. *Consciousness Explained*. Boston: Back Bay Books, 1992.

———. "Philosophy as Naive Anthropology." In *Neuroscience and Philosophy: Brain, Mind, and Language*, by Maxwell Bennett, Daniel Dennett, Peter Hacker, and John Searle, 73–96. New York: Columbia University Press, 2009.

Derrida, Jacques. *Limited Inc*. Edited by Gerald Graff. Translated by Jeffrey Mehlman and Samuel Weber. Evanston: Northwestern University Press, 1988.

———. *Psyche: Inventions of the Other, Volume II*. Edited by Peggy Kamuf and Elizabeth G. Rottenberg. Stanford: Stanford University Press, 2008.

DiCaglio, Joshua. "Scale Tricks and God Tricks, or The Power of Scale in *Powers of Ten*." *Configurations* 28, no. 4 (2020): 459–90. https://doi.org/10.1353/con.2020.0021.

———. "Simulations of Moksha: Liberation, Mysticism, and Transhumanism in Philip K. Dick's *Exegesis*." *Science Fiction Studies* 48 (July 2021): 279–305.

DiCaglio, Joshua, Kathryn M. Barlow, and Joseph S. Johnson. "Rhetorical Recommendations Built on Ecological Experience: A Reassessment of the

Challenge of Environmental Communication." *Environmental Communication* 12, no. 4 (May 19, 2018): 438–50. https://doi.org/10.1080/17524032.2017.1397043.

Dick, Philip K. *The Divine Invasion*. Boston: Mariner Books, 2011.

———. "Exegesis Manuscripts." Online Archive. Zebrapedia Collaborative Transcription Project of Philip K. Dick's Exegesis. zebrapedia.psu.edu.

———. *The Exegesis of Philip K. Dick*. Edited by Pamela Jackson and Jonathan Lethem. Boston: Houghton Mifflin Harcourt, 2011.

———. "How to Build a Universe That Doesn't Fall Apart Two Days Later." In *The Shifting Realities of Philip K. Dick: Selected Literary and Philosophical Writings*, edited by Lawrence Sutin, 259–80. New York: Vintage, 1996.

———. "The Indefatigable Frog." In *Beyond Lies the Wub*, 221–31. London: Gollancz, 1987.

———. *Ubik*. Boston: Mariner Books, 2012.

———. *VALIS*. Boston: Mariner Books, 2011.

Di Palma, Vittoria. "Zoom: Google Earth and Global Intimacy." In *Intimate Metropolis: Urban Subjects in the Modern City*, edited by Vittoria Di Palma, Diana Periton, and Marina Lathouri, 239–61. New York: Routledge, 2009.

Dodge, Martin, and Chris Perkins. "The 'View from Nowhere'? Spatial Politics and Cultural Significance of High-Resolution Satellite Imagery." *Geoforum* 40, no. 4 (July 1, 2009): 497–501. https://doi.org/10.1016/j.geoforum.2009.04.011.

Doyle, Richard. *Darwin's Pharmacy: Sex, Plants, and the Evolution of the Noösphere*. Seattle: University of Washington Press, 2011.

———. "Healing with Plant Intelligence: A Report from Ayahuasca." *Anthropology of Consciousness* 23, no. 1 (Spring 2012): 28–43.

———. *On Beyond Living: Rhetorical Transformations of the Life Sciences*. Stanford: Stanford University Press, 1997.

———. *Wetwares: Experiments in Postvital Living*. Minneapolis: University of Minnesota Press, 2003.

Doyle, Richard, and Trey Connor. "Chanting the Supreme Word of Information: 'Sacred!? Redundant?!'" In *Responding to the Sacred: An Inquiry into the Limits of Rhetoric*, edited by Michael Bernard-Donals and Kyle Jensen, 119–38. University Park: Penn State University Press, 2021.

Drexler, K. Eric. *Engines of Creation: The Coming Era of Nanotechnology*. New York: Anchor, 1987.

Eames, Ray, and Charles Eames. *Powers of Ten: A Film about the Relative Size of Things in the Universe and the Effect of Adding Another Zero*. Los Angeles: IBM/Office of Ray and Charles Eames, 1977. www.powersof10.com/film.

Emerson, Ralph Waldo. "Nature." In *Emerson: Essays and Lectures,* 5–49. New York: Library of America, 1983.

Fahnestock, Jeanne. *Rhetorical Figures of Science.* New York: Oxford University Press, 1999.

Farb, Norman A. S., Zindel V. Segal, Helen Mayberg, Jim Bean, Deborah McKeon, Zainab Fatima, and Adam K. Anderson. "Attending to the Present: Mindfulness Meditation Reveals Distinct Neural Modes of Self-Reference." *Social Cognitive and Affective Neuroscience* 2, no. 4 (December 2007): 313–22. https://doi.org/10.1093/scan/nsm030.

Felski, Rita. *The Limits of Critique.* Chicago: University of Chicago Press, 2015.

Feynman, Richard P. "There's Plenty of Room at the Bottom." *Engineering and Science* 23 (February 1, 1960): 22–36.

Floridi, Luciano. *The Philosophy of Information.* Oxford: Oxford University Press, 2011.

Foerster, Heinz von. *The Beginning of Heaven and Earth Has No Name: Seven Days with Second-Order Cybernetics.* Edited by Albert Müller and Karl H. Müller. Translated by Elinor Rooks and Michael Kasenbacher. New York: Fordham University Press, 2013.

Forman, Robert K. C., ed. *The Problem of Pure Consciousness: Mysticism and Philosophy.* New York: Oxford University Press, 1997.

Foucault, Michel. *The Birth of the Clinic: An Archaeology of Medical Perception.* New York: Vintage, 1994.

———. *The Hermeneutics of the Subject: Lectures at the Collège de France, 1981–1982.* Translated by Graham Burchell. New York: Picador, 2005.

———. *Technologies of the Self: A Seminar with Michel Foucault.* Edited by Luther H. Martin, Huck Gutman, and Patrick H. Hutton. Amherst: University of Massachusetts Press, 1988.

Fox, Kieran C. R., Matthew L. Dixon, Savannah Nijeboer, Manesh Girn, James L. Floman, Michael Lifshitz, Melissa Ellamil, Peter Sedlmeier, and Kalina Christoff. "Functional Neuroanatomy of Meditation: A Review and Meta-analysis of 78 Functional Neuroimaging Investigations." *Neuroscience & Biobehavioral Reviews* 65 (June 1, 2016): 208–28. https://doi.org/10.1016/j.neubiorev.2016.03.021.

Fox, Michael D., Abraham Z. Snyder, Justin L. Vincent, Maurizio Corbetta, David C. Van Essen, and Marcus E. Raichle. "The Human Brain Is Intrinsically Organized into Dynamic, Anticorrelated Functional Networks." *Proceedings of the National Academy of Sciences* 102, no. 27 (July 5, 2005): 9673–78. https://doi.org/10.1073/pnas.0504136102.

Fraser, Nancy. *Scales of Justice: Reimagining Political Space in a Globalizing World.* New York: Columbia University Press, 2010.

Galison, Peter. "The Ontology of the Enemy: Norbert Wiener and the Cybernetic Vision." *Critical Inquiry* 21, no. 1 (1994): 228–66.

Gallop, David. *Parmenides of Elea: A Text and Translation with an Introduction.* Toronto: University of Toronto Press, 1991.

Gamow, George. *One Two Three . . . Infinity: Facts and Speculations of Science.* New York: Dover, 1988.

Gaonkar, Dilip Parameshwar. "The Idea of Rhetoric in the Rhetoric of Science." In *Rhetorical Hermeneutics: Invention and Interpretation in the Age of Science,* edited by Alan G. Gross and William M. Keith, 25–87. Albany: SUNY Press, 1996.

Garb, Yaakov Jerome. "The Use and Misuse of the Whole Earth Image." *Whole Earth Review,* no. 45 (February 4, 1985): 18–25.

Garrison, Kathleen A., Dustin Scheinost, Patrick D. Worhunsky, Hani M. Elwafi, Thomas A. Thornhill, Evan Thompson, Clifford Saron, et al. "Real-Time fMRI Links Subjective Experience with Brain Activity during Focused Attention." *NeuroImage* 81 (November 1, 2013): 110–18. https://doi.org/10.1016/j.neuroimage.2013.05.030.

Geoghegan, Bernard Dionysius. "From Information Theory to French Theory: Jakobson, Lévi-Strauss, and the Cybernetic Apparatus." *Critical Inquiry* 38, no. 1 (2011): 96–126. https://doi.org/10.1086/661645.

Geppert, Alexander C. T. "Rethinking the Space Age: Astroculture and Technoscience." *History and Technology* 28, no. 3 (September 1, 2012): 219–23. https://doi.org/10.1080/07341512.2012.722789.

Gilbert, Scott F., Jan Sapp, and Alfred I. Tauber. "A Symbiotic View of Life: We Have Never Been Individuals." *Quarterly Review of Biology* 87, no. 4 (December 1, 2012): 325–41. https://doi.org/10.1086/668166.

Goleman, Daniel, and Richard J. Davidson. *Altered Traits: Science Reveals How Meditation Changes Your Mind, Brain, and Body*. New York: Avery, 2018.

Graham, S. Scott. *The Politics of Pain Medicine: A Rhetorical-Ontological Inquiry*. Chicago: University of Chicago Press, 2015.

Graner, John, Terrence R. Oakes, Louis M. French, and Gerard Riedy. "Functional MRI in the Investigation of Blast-Related Traumatic Brain Injury." *Frontiers in Neurology* 4 (2013): 16. https://doi.org/10.3389/fneur.2013.00016.

Gross, Alan G. *The Rhetoric of Science.* Cambridge: Harvard University Press, 1990.

Gupta, Bina. *Cit: Consciousness.* New York: Oxford University Press, 2003.

Gurevitch, Leon. "The Digital Globe as Climatic Coming Attraction: From Theatrical Release to Theatre of War." *Canadian Journal of Communication* 38, no. 3 (September 14, 2013): 333–56.

Hacking, Ian. *Representing and Intervening: Introductory Topics in the Philosophy of Natural Science.* Cambridge: Cambridge University Press, 1983.

———. *The Social Construction of What?* Cambridge: Harvard University Press, 1999.

Haldane, J. B. S. "On Being the Right Size." In *Possible Worlds*, 18–26. New York: Routledge, 2002.

Hallward, Peter. *Out of This World: Deleuze and the Philosophy of Creation.* London: Verso, 2006.

Hamilton, J. Paul, Daniella J. Furman, Catie Chang, Moriah E. Thomason, Emily Dennis, and Ian H. Gotlib. "Default-Mode and Task-Positive Network Activity in Major Depressive Disorder: Implications for Adaptive and Maladaptive Rumination." *Biological Psychiatry* 70, no. 4 (August 15, 2011): 327–33. https://doi.org/10.1016/j.biopsych.2011.02.003.

Hanson, Valerie L. *Haptic Visions: Rhetorics of the Digital Image, Information, and Nanotechnology.* Anderson, S.C.: Parlor Press, 2015.

Haraway, Donna. "A Cyborg Manifesto: Science, Technology, and Socialist-Feminism in the Late Twentieth Century." In *Simians, Cyborgs, and Women: The Reinvention of Nature*, 149–82. New York: Routledge, 1990.

———. *Simians, Cyborgs, and Women: The Reinvention of Nature.* New York: Routledge, 1990.

———. "Situated Knowledges: The Science Question in Feminism and the Privilege of Partial Perspective." *Feminist Studies* 14, no. 3 (October 1, 1988): 575–99.

———. *Staying with the Trouble: Making Kin in the Chthulucene.* Durham: Duke University Press Books, 2016.

Harding, Sandra. *The Science Question in Feminism.* Ithaca: Cornell University Press, 1986.

Harrington, Anne. *Reenchanted Science.* Princeton: Princeton University Press, 1999.

Harris, Henry. *The Birth of the Cell.* New Haven: Yale University Press, 1999.

Harris, Randy. "Alan Gross and the Rhetoric of Science." *Perspectives on Science* 17, no. 3 (August 14, 2009): 346–80.

Hathaway, Bill. "Tuning Out: How Brains Benefit from Meditation." *YaleNews*, November 21, 2011.

Hayles, N. Katherine. *How We Became Posthuman: Virtual Bodies in Cybernetics, Literature, and Informatics.* Chicago: University of Chicago Press, 1999.

Heidegger, Martin. *The Question concerning Technology, and Other Essays*. New York: Harper Torchbooks, 1977.

Heise, Ursula K. *Sense of Place and Sense of Planet: The Environmental Imagination of the Global*. Oxford: Oxford University Press, 2008.

Helmreich, Stefan. *Alien Ocean: Anthropological Voyages in Microbial Seas*. Berkeley: University of California Press, 2009.

———. "From Spacehip Earth to Google Ocean: Planetary Icons, Indexes, and Infrastructures." *Social Research* 78, no. 4 (Winter 2011): 1211–42.

Henry, Holly, and Amanda Taylor. "Re-thinking Apollo: Envisioning Environmentalism in Space." *Sociological Review* 57, no. 1 (May 1, 2009): 190–203.

Herod, Andrew. *Scale*. New York: Routledge, 2010.

Herod, Andrew, and Melissa W. Wright, eds. *Geographies of Power: Placing Scale*. Malden, Mass.: Wiley-Blackwell, 2002.

Hird, Myra J. *The Origins of Sociable Life: Evolution after Science Studies*. New York: Palgrave Macmillan, 2009.

Hollenback, Jess. *Mysticism: Experience, Response, and Empowerment*. University Park: Penn State University Press, 2007.

Horton, Zach. "Composing a Cosmic View: Three Alternatives for Thinking Scale in the Anthropocene." In *Scale in Literature and Culture*, edited by Michael Tavel Clarke and David Wittenberg, 35–59. New York: Palgrave Macmillan, 2017.

Howitt, Richard. "Scale." In *A Companion to Political Geography*, edited by John Agnew, Katharyne Mitchell, and Gerard Toal, 138–57. Malden, Mass.: Wiley-Blackwell, 2003.

Humboldt, Alexander von. *Cosmos: A Sketch of a Physical Description of the Universe*. Vol. 1. London: G. Bell, 1893. http://archive.org/details/cosmosofph01humbrich.

Hussey, Edward. *The Presocratics*. Indianapolis: Hackett, 1995.

Huxley, Aldous. *The Perennial Philosophy*. New York: Harper Perennial Modern Classics, 2009.

James, William. *Pragmatism: A New Name for Some Old Ways of Thinking*. New York: Longmans Green, 1907.

———. *The Varieties of Religious Experience*. Oxford: Oxford University Press, 2012.

Jameson, Fredric. Foreword. In *The Postmodern Condition: A Report on Knowledge*, by Jean-François Lyotard, vii–xxi. Minneapolis: University of Minnesota Press, 1984.

Jantzen, Grace M. *Power, Gender, and Christian Mysticism*. Cambridge: Cambridge University Press, 1996.

Jasso, John J. "Sympathy for the Devil: The Myth of Plato as the Enemy of Rhetoric." *Rhetorica: A Journal of the History of Rhetoric* 37, no. 4 (November 1, 2019): 351–81.

Jazairy, El Hadi. "Towards a Plastic Conception of Scale." In *New Geographies, 4: Scales of the Earth,* edited by El Hadi Jazairy, 1. Cambridge: Harvard Graduate School of Design, 2011.

Johnson, Jenell. *American Lobotomy: A Rhetorical History*. Ann Arbor: University of Michigan Press, 2014.

Johnston, William, ed. *The Cloud of Unknowing and The Book of Privy Counseling.* New York: Image, 1996.

Julian of Norwich. *Revelations of Divine Love.* Translated by Elizabeth Spearing. New York: Penguin Books, 1999.

Kant, Immanuel. *Critique of Pure Reason.* Translated by Norman Kemp Smith. 2nd ed. Basingstoke: Palgrave Macmillan, 2007.

Kassem, Salma, Alan T. L. Lee, David A. Leigh, Vanesa Marcos, Leoni I. Palmer, and Simone Pisano. "Stereodivergent Synthesis with a Programmable Molecular Machine." *Nature* 549, no. 7672 (September 2017): 374–78. https://doi.org/10.1038/nature23677.

Katz, Steven T. "Language, Epistemology, and Mysticism." In *Mysticism and Philosophical Analysis,* edited by Katz, 22–74. New York: Oxford University Press, 1978.

Kauffman, Stuart. *At Home in the Universe: The Search for the Laws of Self-Organization and Complexity.* New York: Oxford University Press, 1996.

Keller, Evelyn Fox. *The Mirage of a Space between Nature and Nurture.* Durham: Duke University Press, 2010.

Keller, Evelyn Fox, and Christine R. Grontowski. "The Mind's Eye." In *Discovering Reality: Feminist Perspectives on Epistemology, Metaphysics, Methodology, and Philosophy of Science,* edited by Sandra Harding and Merrill B. Hintikka, 207–24. Dordrecht: D. Reidel, 1983.

Kelsey, Robin. "Reverse Shot: Earthrise and Blue Marble in the American Imagination." In *New Geographies, 4: Scales of the Earth,* edited by El Hadi Jazairy, 11–16. Cambridge: Harvard Graduate School of Design, 2011.

Kierkegaard, Søren. *Fear and Trembling and Sickness unto Death.* Translated by Walter Lowrie. Princeton: Princeton University Press, 2013.

King, Richard. *Orientalism and Religion.* New York: Routledge, 1999.

Kirby, Vicki. *What If Culture Was Nature All Along?* Edinburgh: Edinburgh University Press, 2017.

Kuhn, Thomas S. *The Structure of Scientific Revolutions.* 50th Anniversary Edition. Chicago: University of Chicago Press, 2012.

Kurgan, Laura. *Close Up at a Distance: Mapping, Technology, and Politics.* Brooklyn: Zone Books, 2013.

"Kūṭastha." In *Monier Williams Online Dictionary.* Accessed December 1, 2019. https://www.sanskrit-lexicon.uni-koeln.de/monier/.

Lakoff, George, and Mark Johnson. *Metaphors We Live By.* Chicago: University of Chicago Press, 2003.

Lam, Nina Siu-Ngan, and Dale A. Quattrochi. "On the Issues of Scale, Resolution, and Fractal Analysis in the Mapping Sciences." *Professional Geographer* 44, no. 1 (1992): 88–98.

Latour, Bruno. "Anti-zoom." In *Scale in Literature and Culture,* edited by Michael Tavel Clarke and David Wittenberg, 93–104. New York: Palgrave Macmillan, 2017.

———. "An Attempt at a 'Compositionist Manifesto.'" *New Literary History* 41, no. 3 (2010): 471–90.

———. "Coming Out as a Philosopher." *Social Studies of Science* 40, no. 4 (August 2010): 599–608.

———. "For David Bloor . . . and Beyond: A Reply to David Bloor's 'Anti-Latour.'" *Studies in History and Philosophy of Science Part A* 30, no. 1 (1999): 113–29.

———. *An Inquiry into Modes of Existence: An Anthropology of the Moderns.* Translated by Catherine Porter. Cambridge: Harvard University Press, 2018.

———. "One More Turn after the Social Turn . . ." In *The Science Studies Reader,* edited by Mario Biagioli, 258–76. New York: Routledge, 1999.

———. *Pandora's Hope: Essays on the Reality of Science Studies.* Cambridge: Harvard University Press, 1999.

———. *Politics of Nature: How to Bring the Sciences into Democracy.* Translated by Catherine Porter. Cambridge: Harvard University Press, 2004.

———. *Reassembling the Social: An Introduction to Actor-Network-Theory.* Oxford: Oxford University Press, 2007.

———. *Science in Action.* Cambridge: Harvard University Press, 1987.

———. *We Have Never Been Modern.* Translated by Catherine Porter. Cambridge: Harvard University Press, 1993.

———. "Why Has Critique Run out of Steam? From Matters of Fact to Matters of Concern." *Critical Inquiry* 30, no. 2 (2004): 225–48. https://doi.org/10.1086/421123.

Latour, Bruno, and Steve Woolgar. *Laboratory Life: The Construction of Scientific Facts.* 2nd ed. Princeton: Princeton University Press, 1986.

Lazier, Benjamin. "Earthrise; or, The Globalization of the World Picture." *The American Historical Review* 116, no. 3 (June 1, 2011): 602–30. https://doi.org/10.1086/ahr.116.3.602.

Leonard, Doroethy B. *Franklin Merrell-Wolff: An American Philosopher and Mystic.* Philadelphia: Xlibris, 2016.

Leonard, Ron. *The Transcendental Philosophy of Franklin Merrell-Wolff.* Albany: SUNY Press, 1999.

Levin, Simon A. "The Problem of Pattern and Scale in Ecology." *Ecology* 73, no. 6 (December 1, 1992): 1943–67. https://doi.org/10.2307/1941447.

Li, Wanqing, Xiaoqin Mai, and Chao Liu. "The Default Mode Network and Social Understanding of Others: What Do Brain Connectivity Studies Tell Us." *Frontiers in Human Neuroscience* 8 (2014). https://doi.org/10.3389/fnhum.2014.00074.

Libet, Benjamin. "Unconscious Cerebral Initiative and the Role of Conscious Will in Voluntary Action." *Behavioral and Brain Sciences* 8, no. 4 (December 1985): 529–39. https://doi.org/10.1017/S0140525X00044903.

Lira, Ignacio. "Resolution Revisited." *Metrologia* 43, no. 3 (March 2006): L14–L17. https://doi.org/10.1088/0026-1394/43/3/N03.

Littlefield, Melissa M., and Jenell M. Johnson, eds. *The Neuroscientific Turn: Transdisciplinarity in the Age of the Brain.* Preface by Judy Illes. Ann Arbor: University of Michigan Press, 2012.

Lovejoy, Arthur O. *The Great Chain of Being: A Study of the History of an Idea.* Rev. ed. Cambridge: Harvard University Press, 1976.

Lucian of Samosata. *Selected Satires of Lucian.* Translated by Lionel Casson. New York: Norton, 1968.

Luhmann, Niklas. *Introduction to Systems Theory.* Translated by Peter Gilgen. Cambridge, UK: Polity, 2012.

———. *Theories of Distinction: Redescribing the Descriptions of Modernity.* Edited by William Rasch. Stanford: Stanford University Press, 2002.

Lukinbeal, Christopher. "Scale: An Unstable Representational Analogy." *Media Fields Journal,* no. 4 (2012): 1–13.

Macpherson, Sandra, and Meredith Martin. "Essays from the English Institute 2017: Scale." *ELH* 86, no. 2 (June 4, 2019): 267–74. https://doi.org/10.1353/elh.2019.0014.

Maharshi, Ramana. *Be as You Are: The Teachings of Sri Ramana Maharshi.* Edited by David Godman. New Delhi: Penguin, 1988.

———. *Who Am I? The Teachings of Bhagavan Sri Ramana Maharshi.* 24th ed. Tiruvannamalai: Sri Ramanasramam, 2008.

Mandelbrot, Benoit. *The Fractal Geometry of Nature.* San Francisco: W. H. Freeman, 1983.

———. *The Fractalist: Memoir of a Scientific Maverick.* New York: Vintage, 2012.

Mansfield, Nick. *Subjectivity: Theories of the Self from Freud to Haraway*. New York: NYU Press, 2000.

Margulis, Lynn. "Symbiogenesis: A New Principle of Evolution Rediscovery of Boris Mikhaylovich Kozo-Polyansky (1890–1957)." *Paleontological Journal* 44, no. 12 (December 1, 2010): 1525–39. https://doi.org/10.1134/S0031030110120087.

———. *Symbiotic Planet: A New Look at Evolution*. New York: Basic Books, 1999.

Margulis, Lynn, and Dorion Sagan. *Dazzle Gradually: Reflections on the Nature of Nature*. Reprint. White River Junction, Vt.: Chelsea Green, 2007.

———. *Microcosmos*. Berkeley: University of California Press, 1997.

———. *What Is Life?* Berkeley: University of California Press, 2000.

Marr, David. *Vision: A Computational Investigation into the Human Representation and Processing of Visual Information*. Cambridge: MIT Press, 2010.

Marston, Sallie A. "The Social Construction of Scale." *Progress in Human Geography* 24, no. 2 (June 1, 2000): 219–42. https://doi.org/10.1191/030913200674086272.

Marston, Sallie A., John Paul Jones, and Keith Woodward. "Human Geography without Scale." *Transactions of the Institute of British Geographers* 30, no. 4 (December 1, 2005): 416–32. https://doi.org/10.1111/j.1475-5661.2005.00180.x.

Massumi, Brian. *Parables for the Virtual: Movement, Affect, Sensation*. Durham: Duke University Press, 2002.

Maturana, Humberto R. Introduction. In *Autopoiesis and Cognition: The Realization of the Living*, by Humberto R. Maturana and Francisco J. Varela, xi–xxx. Boston: D. Reidel, 1972.

Maturana, Humberto R., and Francisco J. Varela. *Autopoiesis and Cognition: The Realization of the Living*. Boston: D. Reidel, 1972.

Mazzarello, Paolo. "A Unifying Concept: The History of Cell Theory." *Nature Cell Biology* 1, no. 1 (May 1999): E13–E15. https://doi.org/10.1038/8964.

McGuirk, Kevin. "A. R. Ammons and the Whole Earth." *Cultural Critique*, no. 37 (1997): 131–58. https://doi.org/10.2307/1354543.

Merleau-Ponty, Maurice. *Phenomenology of Perception*. 2nd ed. London: Routledge, 2002.

Merchant, Carolyn. *Reinventing Eden: The Fate of Nature in Western Culture*. London: Routledge, 2004.

Merrell-Wolff, Franklin. *Franklin Merrell-Wolff's Experience and Philosophy: A Personal Record of Transformation and a Discussion of Transcendental Consciousness: Containing His Philosophy of Consciousness without an Object and His Pathways through to Space*. Albany: SUNY Press, 1994.

———. *Transformations in Consciousness: The Metaphysics and Epistemology.* Albany: SUNY Press, 1995.

Messeri, Lisa. *Placing Outer Space: An Earthly Ethnography of Other Worlds.* Durham: Duke University Press, 2016.

Milburn, Colin. *Mondo Nano: Fun and Games in the World of Digital Matter.* Durham: Duke University Press, 2015.

———. *Nanovision: Engineering the Future.* Durham: Duke University Press, 2008.

Mitchell, Edgar. *The Way of the Explorer: An Apollo Astronaut's Journey through the Material and Mystical Worlds.* Rev. ed. Franklin Lakes, N.J.: New Page Books, 2008.

Mol, Annemarie. *The Body Multiple: Ontology in Medical Practice.* Durham: Duke University Press, 2003.

Morton, Timothy. *The Ecological Thought.* Cambridge: Harvard University Press, 2012.

———. *Ecology without Nature: Rethinking Environmental Aesthetics.* Cambridge: Harvard University Press, 2009.

———. *Hyperobjects: Philosophy and Ecology after the End of the World.* Minneapolis: University of Minnesota Press, 2013.

Mulvey, Laura. "Visual Pleasure and Narrative Cinema." *Screen* 16, no. 3 (September 21, 1975): 6–18. https://doi.org/10.1093/screen/16.3.6.

Myers, Natasha. *Rendering Life Molecular: Models, Modelers, and Excitable Matter.* Durham: Duke University Press, 2015.

Nagarjuna. *Nagarjuna's Middle Way: Mūlamadhyamakakārikā.* Translated by Mark Siderits and Shoryu Katsura. Boston: Wisdom Publications, 2013.

Nagel, Thomas. *The View from Nowhere.* New York: Oxford University Press, 1989.

———. "What Is It Like to Be a Bat?" *Philosophical Review* 83, no. 4 (October 1974): 435–50. https://doi.org/10.2307/2183914.

Nealon, Jeffrey. *Post-postmodernism: or, The Cultural Logic of Just-in-Time Capitalism.* Stanford: Stanford University Press, 2012.

Nicholas of Cusa. *The Vision of God.* Translated by Emma Gurney Salter. New York: Cosimo Classics, 2007.

Nolan, Christopher, dir. *Interstellar*. Paramount Pictures, 2014.

Olivelle, Patrick, ed. *The Early Upanishads: Annotated Text and Translation.* New York: Oxford University Press, 1998.

Olwig, Kenneth R. "The Earth Is Not a Globe: Landscape versus the 'Globalist' Agenda." *Landscape Research* 36, no. 4 (August 1, 2011): 401–15. https://doi.org/10.1080/01426397.2011.582940.

O'Neill, R. V., and A. W. King. "Homage to St. Michael; or, Why Are There So Many Books on Scale?" In *Ecological Scale*, edited by David Lawrence Peterson and V. Thomas Parker, 3–16. New York: Columbia University Press, 1998.

O'Neill, Saffron J., and Nicholas Smith. "Climate Change and Visual Imagery." *WIREs Climate Change* 5, no. 1 (2014): 73–87. https://doi.org/10.1002/wcc.249.

Ouspensky, P. D. *Tertium Organum: The Third Canon of Thought*. Translated by Nicholas Bessaraboff and Claude Bragdon. New York: Knopf, 1922.

Pahnke, Walter N. "The Psychedelic Mystical Experience in the Human Encounter with Death." *Harvard Theological Review* 62, no. 1 (January 1969): 1–21. https://doi.org/10.1017/S0017816000027577.

Pattee, Howard H., ed. *Hierarchy Theory: The Challenge of Complex Systems*. New York: George Braziller, 1973.

———. "How Does a Molecule Become a Message?" In *Developmental Biology Supplement*, edited by Anton Lang, 1–16. New York: Academic Press, 1969.

Peirce, Charles Sanders. "On a New List of Categories." In *Peirce on Signs: Writings on Semiotic by Charles Sanders Peirce*, edited by James Hoopes, 23–33. Chapel Hill: University of North Carolina Press, 1991.

Pickering, Andrew. *The Mangle of Practice: Time, Agency, and Science*. Chicago: University of Chicago Press, 1995.

Pilsch, Andrew. *Transhumanism: Evolutionary Futurism and the Human Technologies of Utopia*. Minneapolis: University of Minnesota Press, 2017.

Plato. *Alcibiades 1*. Translated by W. R. M. Lamb. Vol. 8. *Plato in Twelve Volumes*. Cambridge: Harvard University Press, 1955.

———. *Phaedo*. Translated by Harold Fowler. Vol. 1. *Plato in Twelve Volumes*. Cambridge: Harvard University Press, 1966.

———. *Phaedrus*. In *Plato: Complete Works*, edited by John M. Cooper, translated by Alexander Nehamas and Paul Woodruff, 506–56. Indianapolis: Hackett, 1997.

———. *Protagoras*. In *Plato: Complete Works*, edited by John M. Cooper, translated by Stanley Lombardo and Karen Bell, 746–90. Indianapolis: Hackett, 1997.

———. "Seventh Letter." In *Plato: Complete Works*, edited by John M. Cooper, translated by Glenn R. Morrow, 1646–67. Indianapolis: Hackett, 1997.

———. *Sophist*. In *Plato: Complete Works*, edited by John M. Cooper, translated by Nicholas P. White, 235–93. Indianapolis: Hackett, 1997.

———. *Theaetetus*. Translated by Harold Fowler. Vol. 12. *Plato in Twelve Volumes*. Cambridge: Harvard University Press, 1921.

——. *Timaeus*. In *Plato: Complete Works,* edited by John M. Cooper, translated by Donald J. Zeyl, 1224–91. Indianapolis: Hackett, 1997.

Plotinus. *The Enneads.* Translated by Stephen MacKenna. Rev. ed. Burdett, N.Y.: Larson, 1992.

Polanyi, Michael. *The Tacit Dimension.* Chicago: University of Chicago Press, 2009.

Poole, Robert. *Earthrise: How Man First Saw the Earth.* New Haven: Yale University Press, 2008.

Post, David M., Martin W. Doyle, John L. Sabo, and Jacques C. Finlay. "The Problem of Boundaries in Defining Ecosystems: A Potential Landmine for Uniting Geomorphology and Ecology." *Geomorphology,* 36th Binghamton Geomorphology Symposium, 89, no. 1 (September 1, 2007): 111–26. https://doi.org/10.1016/j.geomorph.2006.07.014.

Prigogine, Ilya, and Isabelle Stengers. *Order out of Chaos: Man's New Dialogue with Nature.* London: Verso, 2018.

Primack, Joel R., and Nancy Ellen Abrams. *The View from the Center of the Universe: Discovering Our Extraordinary Place in the Cosmos.* New York: Riverhead Books, 2007.

Pseudo-Dionysius. *Pseudo-Dionysius: The Complete Works.* Translated by Paul Rorem. New York: Paulist Press, 1988.

Rai, Candice, and Caroline Gottschalk Druschke, eds. *Field Rhetoric: Ethnography, Ecology, and Engagement in the Places of Rhetoric.* Tuscaloosa: University of Alabama Press, 2018.

Raichle, M. E., A. M. MacLeod, A. Z. Snyder, W. J. Powers, D. A. Gusnard, and G. L. Shulman. "A Default Mode of Brain Function." *Proceedings of the National Academy of Sciences* 98, no. 2 (January 16, 2001): 676–82. https://doi.org/10.1073/pnas.98.2.676.

Rao, Lingeswar, and Anil Sharma, trans. *The Ribhu Gita (Sixth Amsa of Siva Rahasyam).* 2nd ed. Tiruvannamalai, India: Sri Ramnasramam, 2012.

Rheinberger, Hans-Jörg. *Toward a History of Epistemic Things: Synthesizing Proteins in the Test Tube.* Stanford: Stanford University Press, 1997.

Roberts, Dorothy. *Killing the Black Body: Race, Reproduction, and the Meaning of Liberty.* New York: Vintage, 1998.

Robinson, T. M., trans. *Heraclitus: Fragments.* Toronto: University of Toronto Press, 1987.

Rotman, Brian. *Ad Infinitum: The Ghost in Turing's Machine; Taking God out of Mathematics and Putting the Body Back In.* Stanford: Stanford University Press, 1993.

——. *Becoming Beside Ourselves: The Alphabet, Ghosts, and Distributed Human Being*. Durham: Duke University Press Books, 2008.

——. *Signifying Nothing: The Semiotics of Zero.* Stanford: Stanford University Press, 1993.

Russell, Bertrand. "Mysticism and Logic." In *Mysticism and Logic and Other Essays,* 1–32. London: Longmans, Green, 1918.

Russill, Chris. "Earth-Observing Media." *Canadian Journal of Communication* 38, no. 3 (September 14, 2013). http://www.cjc-online.ca/index.php/journal/article/view/2756.

Ryan, Joel, Abby R. Gerhold, Vincent Boudreau, Lydia Smith, and Paul S. Maddox. "Introduction to Modern Methods in Light Microscopy." In *Light Microscopy: Methods and Protocols,* edited by Yolanda Markaki and Hartmann Harz, 1–15. New York: Springer, 2017. https://doi.org/10.1007/978-1-4939-6810-7_1.

Ryle, Gilbert. *The Concept of Mind.* 60th Anniversary Edition. New York: Routledge, 2009.

Sagan, Carl. *Billions and Billions: Thoughts on Life and Death at the Brink of the Millennium.* New York: Random House, 1997.

——, narrator. *Cosmos: A Personal Voyage.* Episode 1, "Shores of the Cosmic Ocean." Created by Ann Druyan, Carl Sagan, and Steven Soter. Aired September 28, 1980, on Public Broadcasting Service.

——. *The Demon-Haunted World: Science as a Candle in the Dark.* New York: Ballantine Books. 1996.

——. *The Varieties of Scientific Experience: A Personal View of the Search for God.* Edited by Ann Druyan. Penguin Books, 2007.

Salthe, Stanley, and Gary Fuhrman. "The Cosmic Bellows: The Big Bang and the Second Law." *Cosmos and History: The Journal of Natural and Social Philosophy* 1, no. 2 (January 19, 2006): 295–318.

Sapp, Jan. *The New Foundations of Evolution: On the Tree of Life.* Oxford: Oxford University Press, 2009.

Sargeant, Winthrop, ed. *The Bhagavad Gita.* Twenty-Fifth-Anniversary Bilingual Edition. Albany: Excelsior Editions, 2009.

Schiappa, Edward. "Did Plato Coin *Rhētorikē*?" *American Journal of Philology* 111, no. 4 (1990): 457–70. https://doi.org/10.2307/295241.

Schickore, Jutta. *The Microscope and the Eye: A History of Reflections, 1740–1870.* Chicago: University of Chicago Press, 2007.

Schrodinger, Erwin. *My View of the World.* Translated by Cecily Hastings. Cambridge: Cambridge University Press, 1964.

Schweickart, Rusty. "No Frames, No Boundaries." Audio recording. Lindisfarne, 1974. Lindisfarne Tapes—Schumacher Center for New Economics. https://centerforneweconomics.org/envision/legacy/lindisfarne-tapes/.

Shankara. *Direct Awareness of the Self: A Translation of the Aparokṣānubhūti by Sankara.* Translated by Douglas A. Fox. Lewiston, N.Y.: Edwin Mellen Press, 1995.

———. *The Vivekacūdāmani of Shankarācārya Bhagavatpāda: An Introduction and Translation.* Translated and edited by John Grimes. Aldershot: Ashgate Publishing, 2004.

Shannon, Claude Elwood, and Warren Weaver. *The Mathematical Theory of Communication.* Urbana: University of Illinois Press, 1964.

Shapin, Steven, and Simon Schaffer. *Leviathan and the Air-Pump: Hobbes, Boyle, and the Experimental Life.* 1985. Princeton: Princeton University Press, 2011.

Sheline, Yvette I., Deanna M. Barch, Joseph L. Price, Melissa M. Rundle, S. Neil Vaishnavi, Abraham Z. Snyder, Mark A. Mintun, Suzhi Wang, Rebecca S. Coalson, and Marcus E. Raichle. "The Default Mode Network and Self-Referential Processes in Depression." *Proceedings of the National Academy of Sciences of the United States of America* 106, no. 6 (February 10, 2009): 1942–47. https://doi.org/10.1073/pnas.0812686106.

Smalley, Richard. "Of Chemistry, Love, and Nanobots." *Scientific American,* September 2001, 76–77. https://doi.org/10.1038/scientificamerican0901-76.

Solé, Ricard. *Phase Transitions.* Princeton: Princeton University Press, 2011.

Sonea, Sorin, and Maurice Panisset. *A New Bacteriology.* Boston: Jones and Bartlett Learning, 1983.

Sontag, Susan. *On Photography.* New York: Picador, 2001.

Soon, Chun Siong, Marcel Brass, Hans-Jochen Heinze, and John-Dylan Haynes. "Unconscious Determinants of Free Decisions in the Human Brain." *Nature Neuroscience* 11, no. 5 (May 2008): 543–45. https://doi.org/10.1038/nn.2112.

Stace, W. T. *Mysticism and Philosophy.* Los Angeles: Jeremy P. Tarcher, 1987.

Stone, A. Douglas. *Einstein and the Quantum: The Quest of the Valiant Swabian.* Princeton: Princeton University Press, 2015.

Szasz, Eva, dir. *Cosmic Zoom.* 1968. 8:02.

Tang, Yi-Yuan, Britta K. Hölzel, and Michael I. Posner. "The Neuroscience of Mindfulness Meditation." *Nature Reviews Neuroscience* 16, no. 4 (April 2015): 213–25. https://doi.org/10.1038/nrn3916.

Taylor, Jill Bolte. "My Stroke of Insight." TED Talks, February 2008. https://www.ted.com/. 17:17.

Taylor, Véronique A., Véronique Daneault, Joshua Grant, Geneviève Scavone, Estelle Breton, Sébastien Roffe-Vidal, Jérômfe Courtemanche, et al. "Impact of Meditation Training on the Default Mode Network during a Restful State." *Social Cognitive and Affective Neuroscience* 8, no. 1 (January 1, 2013): 4–14. https://doi.org/10.1093/scan/nsr087.

Teilhard de Chardin, Pierre. *The Phenomenon of Man.* New York: Harper Perennial Modern Classics, 2008.

Teston, Christa. *Bodies in Flux: Scientific Methods for Negotiating Medical Uncertainty*. Chicago: University of Chicago Press, 2017.

Thompson, D'Arcy Wentworth. *On Growth and Form: The Complete Revised Edition.* New York: Dover, 1992.

Tomes, Nancy. *The Gospel of Germs: Men, Women, and the Microbe in American Life*. Cambridge: Harvard University Press, 1998.

Tong, Chris. "Ecology without Scale: Unthinking the World Zoom." *Animation* 9, no. 2 (July 1, 2014): 196–211. https://doi.org/10.1177/1746847714527199.

Tribbe, Matthew D. *No Requiem for the Space Age: The Apollo Moon Landings and American Culture.* Oxford: Oxford University Press, 2014.

Tsing, Anna Lowenhaupt. *The Mushroom at the End of the World: On the Possibility of Life in Capitalist Ruins.* Princeton: Princeton University Press, 2017.

———. "On Nonscalability: The Living World Is Not Amenable to Precision-Nested Scales." *Common Knowledge* 18, no. 3 (August 1, 2012): 505–24. https://doi.org/10.1215/0961754X-1630424.

Tung, Charles M. "Baddest Modernism: The Scales and Lines of Inhuman Time." *Modernism/Modernity* 23, no. 3 (September 27, 2016): 515–38. https://doi.org/10.1353/mod.2016.0049.

Uddin, Lucina Q., A. M. Kelly, Bharat B. Biswal, F. Xavier Castellanos, and Michael P. Milham. "Functional Connectivity of Default Mode Network Components: Correlation, Anticorrelation, and Causality." *Human Brain Mapping* 30, no. 2 (February 2009): 625–37. https://doi.org/10.1002/hbm.20531.

Uexküll, Jakob von. *A Foray into the Worlds of Animals and Humans: With a Theory of Meaning.* Minneapolis: University of Minnesota Press, 2013.

Underhill, Evelyn. *Mysticism: A Study in the Nature and Development of Spiritual Consciousness.* Mineola, N.Y.: Dover, 2002.

Vadde, Aarthi. "Scalability." Edited by Thomas Davis and Nathan Hensley. *Modernism/Modernity Print Plus* 2, no. 4 (January 4, 2018). https://doi.org/10.26597/mod.0035.

"Vajracchedikā Prajñāpāramitā." Translated by Paul Harrison. Bibliotheca Polyglotta. Accessed December 3, 2019. https://www2.hf.uio.no/polyglotta/index.php?page=fulltext&vid=22&view=fulltext.

Van Dam, Nicholas T., Marieke K. van Vugt, David R. Vago, Laura Schmalzl, Clifford D. Saron, Andrew Olendzki, Ted Meissner, et al. "Mind the Hype:

A Critical Evaluation and Prescriptive Agenda for Research on Mindfulness and Meditation." *Perspectives on Psychological Science: A Journal of the Association for Psychological Science* 13, no. 1 (2018): 36–61. https://doi.org/10.1177/1745691617709589.

Vanhaudenhuyse, Audrey, Athena Demertzi, Manuel Schabus, Quentin Noirhomme, Serge Bredart, Melanie Boly, Christophe Phillips, et al. "Two Distinct Neuronal Networks Mediate the Awareness of Environment and of Self." *Journal of Cognitive Neuroscience* 23, no. 3 (March 2011): 570–78. https://doi.org/10.1162/jocn.2010.21488.

Vernadsky, Vladimir I. "The Biosphere and the Noösphere." *American Scientist* 33, no. 1 (1945): 1–12.

Versluis, Arthur. *Platonic Mysticism: Contemplative Science, Philosophy, Literature, and Art.* Albany: SUNY Press, 2018.

Vieten, Cassandra, Helané Wahbeh, B. Rael Cahn, Katherine MacLean, Mica Estrada, Paul Mills, Michael Murphy, et al. "Future Directions in Meditation Research: Recommendations for Expanding the Field of Contemplative Science." *PLOS ONE* 13, no. 11 (November 7, 2018): e0205740. https://doi.org/10.1371/journal.pone.0205740.

Virno, Paolo. *A Grammar of the Multitude: For an Analysis of Contemporary Forms of Life.* Translated by Isabella Bertoletti, James Cascaito, and Andrea Casson. London: Semiotext(e), 2004.

Vitale, Francesco. *Biodeconstruction: Jacques Derrida and the Life Sciences.* Translated by Mauro Senatore. Albany: SUNY Press.

Walls, Laura Dassow. *The Passage to Cosmos: Alexander von Humboldt and the Shaping of America.* Chicago: University of Chicago Press, 2009.

Walsh, Lynda. "The Visual Rhetoric of Climate Change." *WIREs Climate Change* 6, no. 4 (2015): 361–68. https://doi.org/10.1002/wcc.342.

Walsh, Lynda, Nathaniel A. Rivers, Jenny Rice, Laurie E. Gries, Jennifer L. Bay, Thomas Rickert, and Carolyn R. Miller. "Forum: Bruno Latour on Rhetoric." *Rhetoric Society Quarterly* 47, no. 5 (October 20, 2017): 403–62. https://doi.org/10.1080/02773945.2017.1369822.

Webb, Stephen. *Measuring the Universe: The Cosmological Distance Ladder.* London: Springer, 1999.

Weber, Gary. *Happiness beyond Thought: A Practical Guide to Awakening.* Lincoln, Neb.: iUniverse, 2007.

———. "What 'No Thoughts' Means—3 Different Kinds of Thoughts." YouTube video, July 14, 2013. https://www.youtube.com/watch?v=WnWxCgiZfrc.

Weber, Gary, and Richard Doyle. *Into the Stillness: Dialogues on Awakening beyond Thought.* Salisbury: Non-Duality Press, 2015.

Welling, Bart. "Ecoporn: On the Limits of Visualizing the Nonhuman." In *Ecosee: Image, Rhetoric, Nature,* edited by Sidney I. Dobrin and Sean Morey, 53–77. Albany: SUNY Press, 2009.

Wells, H. G. *The War of the Worlds.* 1898. Reprint. New York: Aerie, 1993.

West, Geoffrey. *Scale: The Universal Laws of Life, Growth, and Death in Organisms, Cities, and Companies.* New York: Penguin Books, 2017.

Wheelwright, Philip. *Heraclitus.* Special Edition. New York: Oxford University Press, 1959.

White, Frank. *The Overview Effect: Space Exploration and Human Evolution.* 3rd ed. Reston, Va.: American Institute of Aeronautics, 2014.

White, William J. "Disciplinarity and the Rhetoric of Science: A Social Epistemological Reception Study." *Poroi* 10, no. 2 (December 30, 2014). https://doi.org/10.13008/2151-2957.1130.

Whitman, Walt. *Democratic Vistas: The Original Edition in Facsimile.* Edited by Ed Folsom. Iowa City: University of Iowa Press, 2009.

———. *Song of Myself.* In *Leaves of Grass.* Philadelphia: David McKay, 1892.

Wiener, Norbert. *The Human Use of Human Beings: Cybernetics and Society.* New York: Da Capo Press, 1988.

Wilson, Elizabeth A. *Gut Feminism.* Durham: Duke University Press, 2015.

———. *Neural Geographies: Feminism and the Microstructure of Cognition.* New York: Routledge, 1998.

Wolfe, Cary. *Critical Environments: Postmodern Theory and the Pragmatics of the "Outside."* Minneapolis: University of Minnesota Press, 1998.

———. *What Is Posthumanism?* Minneapolis: University of Minnesota Press, 2009.

Wolfram, Stephen. *A New Kind of Science.* Champaign, Ill.: Wolfram Media, 2002.

Woods, Derek. "Epistemic Things in Charles and Ray Eames's Powers of Ten." In *Scale in Literature and Culture,* edited by Michael Tavel Clarke and David Wittenberg, 61–92. New York: Palgrave Macmillan, 2017.

———. "Scale Critique for the Anthropocene." *Minnesota Review* 2014, no. 83 (November 1, 2014): 133–42. https://doi.org/10.1215/00265667-2782327.

Worth, Josh. *If the Moon Were Only 1 Pixel—A Tediously Accurate Map of the Solar System.* 2014. https://joshworth.com/dev/pixelspace/pixelspace_solar system.html.

Wright, Robert. *Why Buddhism Is True: The Science and Philosophy of Meditation and Enlightenment.* New York: Simon and Schuster Paperbacks, 2017.

Wu, Jianguo, and Harbin Li. "Concepts of Scale and Scaling." In *Scaling and Uncertainty Analysis in Ecology: Methods and Applications,* edited by Jianguo

Wu, K. Bruce Jones, Harbin Li, and Orie L. Loucks, 1–13. Dordrecht: Springer, 2006.

Young, Stephen S., and Paul Kelly. "Macro or Micro? A Visual Art Exhibition Challenging Our Perceptions of Scale." *GeoHumanities* 3, no. 1 (January 2, 2017): 250–65. https://doi.org/10.1080/2373566X.2017.1295815.

Zylinska, Joanna. *Minimal Ethics for the Anthropocene.* Ann Arbor: Open Humanities Press, 2014.

Index

Page numbers in italic refer to figures.

JOSHUA DICAGLIO is assistant professor of English at Texas A&M University.